D0763995

The secret life of

flies

The secret life of

flies

Erica McAlister

FIREFLY BOOKS

A FIREFLY BOOK

For Alfie (TLB)

Published by Firefly Books Ltd. 2017

Copyright © 2017 The Trustees of the Natural History Museum, London

First printing

Publisher Cataloging-in-Publication Data (U.S.)

A CIP record for this title is available from the Library of Congress

Library and Archives Canada Cataloguing in Publication

A CIP record for this title is available from Library and Archives Canada

Published in the United States by
Firefly Books (U.S.) Inc.
P.O. Box 1338, Ellicott Station
Buffalo, New York 14205

Published in Canada by
Firefly Books Ltd.
50 Staples Avenue, Unit 1
Richmond Hill, Ontario L4B 0A7

Printed in China

First published by the
Natural History Museum,
Cromwell Road,
London SW7 5BD

The Author has asserted her
right to be identified as the
Author of this work under
the Copyright, Designs and
Patents Act 1988

Designed by Bobby Birchall,
Bobby&Co; Reproduction by
Saxon Digital Services

Front cover: *Formosia moneta*, a parasitic fly
Back cover and page 1: *Achias rothschildi*, a stalk-eyed fly
Page 2: *Pegesimallus teratodes*, a robber fly

Contents

sericata
Meigen

sericata
Meigen

sericata
Meigen

Introduction

*All the efforts of the human mind cannot exhaust
the essence of a single fly.*

Thomas Aquinas

F LIES. A nuisance at best, a harbinger of death at worst. Regarded by many as a disease-carrier that vomits on our food, it earns nothing more from us humans than feelings of disgust. The little we know about the fly we don't like. I've seen more people intentionally swat and kill flies than any other creature, and without a second thought. But there is another side to its story. To me, the fly is one of nature's great marvels, and I have been fascinated by them since childhood. I always had a thing for nature and the small stuff, and I wasn't immune to the joys of an odd decomposing corpse I found in the garden. I was intrigued by these wiggling, feeding machines and this early fascination has never left me. Flies, to my mind, are the most complex, crucial and highly adaptive creatures on the planet. Their story reveals how life without them would be no life at all. Can we say that about our own species?

Take time to look at flies, really look at how they're built, and you'll see beauty. They have amazingly diverse forms, bodies

These specimens of the green bottle, *Lucilia sericata*, all look the same but differences between their DNA help us study any changes in their populations.

flamboyant with colourful metallic sheens that play in the light, wings cloaked in pearly veils. The eyes of some species are banded in brilliant colours, while the charmingly fuzzy features of others can disarm even those claiming to be repulsed by flies and lead to them being mistaken for unusual-looking bees. These are but a few of the features that to this day have me squealing like a child when I go out collecting and observing them, or when rummaging through the collections at the Natural History Museum in London.

These little beauties are everywhere. They are found on every continent, and have even ventured where few other insects have gone – the sea. Some species live in the splash zones of beaches and others have properly invaded the marine environment, living just below the surface of the water around tropical atolls. True, to date not many species of fly have been described from this environment, only four in fact, but no other insect lives there. The point is that flies live everywhere. And they don't travel alone.

So let's talk numbers. We estimate that there are approximately 10 quintillion insects alive at any one moment, so around 200 million insects for every human. And about eight and a half per cent of those insects are estimated to be flies, which works out at 17 million flies for each and every one of us. Does that make your skin crawl? And flies make up a large percentage of the total number of described species on the planet – one in 10 of every species described is a fly. But the question you still want answering is: what have all these flies ever done for us?

The impact of flies on our environment is hard to overestimate. They are not just important, they are essential for maintaining healthy ecosystems. They pollinate flowers and, wait for it, if it wasn't for flies pollinating the plant *Theobroma cacao* we would live in a world without chocolate. They are excellent pest controllers, for instance hover fly larvae eat the aphids that destroy our garden blooms, and they are a key food source for many other animals, especially birds. They are also very good at decomposing waste, a job no one else really wants to do, and they can indicate the quality of our water systems. With so many valuable contributions, it's no wonder they are thriving across the globe.

I'll give you an example of how valuable I know flies to be. I recently travelled to central Peru in South America – Huarez to be exact. Fuelled by marvellous lattes and armed with a backpack aspirator (makes me look like a Ghostbuster), as well as nets and

One in ten of every species described so far is a fly. Here, blow flies feast on a fish head.

a hand-held aspirator I spent several weeks collecting flies – I was a sight that caused much amusement to all who happened to be passing. I was looking specifically for flies that were active on or near the wild relatives of Solanaceae, which is the family of plants that contain some of the most important crop species to humans – potatoes and tomatoes. Peru is the birthplace of these crops, and there are more than 200 wild species or subspecies still found growing au naturel high up in the Andes. Nowadays, due to ever-growing human population numbers, we are becoming increasingly concerned about how we are going to feed ourselves. Scientists are examining these wild species of Solanaceae and developing new, and hopefully more robust, cultivars that can produce larger yields, withstand pest epidemics and of course, the big one, survive climate change. Knowing which insects pollinate and predate these crops (flies do both), and how their distribution varies from region to region, is crucial information.

The future of our global food supply relies on our knowledge of which insects pollinate our many plants. But what we know about the impact of flies on this supply is sorely lacking, so we need to get our skates on and get out there, often to inhospitable environments, and describe what we find. Any new species discovered, as well as the described (and undescribed) specimens from museum collections, will hopefully enable us to predict pest and pollinator population change in relation to changes in the host plant range.

Sadly, with many species of insect we have a long way to go in terms of getting to grips with understanding them, and this is especially the case with flies. Mammals and birds suck up all the taxonomic (and more importantly the financial) attention, being seen as the charismatic fauna, and they have a disproportionate number of biologists studying them! I have had many an argument with primatologists (what I call monkey lovers) about what is the point of sitting under a tree waiting for a monkey to do something

The value of this common fruit fly, *Drosophila melanogaster*, in research is immeasurable. It has helped in the understanding of some human disease-causing genes as it shares 75% of such genes with us.

exciting, like defecating, unless you are going to use the faeces as bait to sample for insects. Many believe that primates have more interesting behavioural traits and so are the only useful models for understanding our own societies, but this is simply not the case. Many flies demonstrate complex behaviours, like throwing out incredibly intricate dance moves to impress the opposite sex in a manner similar to many of the larger mammals. In fact, studying the behaviour of flies can help us understand some of the behaviours of our own species and we can do this in some remarkable ways.

One of the best examples is with studies of a very small – it is only about three millimetres (¼ in) long – but very important fly. The common fruit fly, *Drosophila melanogaster* (although it is not a fruit fly in the strictest sense of belonging to the fruit fly family but rather it is part of the vinegar fly family), has over the past 100 years shaped modern genetics. Without the research into this little fly's genetics we would not have such a good understanding of human disease genes, such as those associated with Alzheimer's and Parkinson's.

Deoxyribonucleic acid (DNA) is a long molecule that carries the genetic instructions used in the growth, development, functioning and reproduction of all organisms. Genes are sections of the DNA and each gene has a sequence of 'base pairs'. Although our DNA is much longer than the DNA of *Drosophila melanogaster* (3,200 million base pairs compared with 130 million base pairs), we do not have many more genes – our 20,000 compared with their 13,600 – and 75% of disease-causing genes are shared between this species and us. Its love of breeding, and the ease with which we have been able to switch its genes on and off, make it a lovely model for us to determine the impacts and effects of inherited diseases and of drugs, gene manipulation and environmental stress. It was the first animal to have its genome sequenced back in 2000, which led the way to the sequencing of ours three years later.

And our research using this species has not been wholly earthbound. They were also the first creatures we sent into space, and they are still acting as mini-astronauts. Currently there is a fruit fly laboratory on the International Space Station where research is being undertaken to see how the lack of gravity affects the flies' health and from that scientists can estimate how space travel is likely to affect our own health.

Our lack of knowledge about flies is quite extraordinary when compared to what we know about other groups of animals, even other groups of insects. When Carl Linnaeus, the Swedish taxonomist who devised the scientific naming system we still use to categorize species, published his seminal book on animal classification in 1758, *Systema naturæ per regna tria naturæ, secundum classes, ordines, genera, species, cum characteribus, differentiis, synonymis, locis*, he listed only 191 species of flies in 10 groups, or genera. Linnaeus had a two-name naming system of genus and species, the former being a group of very similar organisms, for example the *Drosophila* genus. Within this genus there are many types of *Drosophila* which are thought not to interbreed and that represent different species. Families represent the next level of grouping. To date we have scientifically recorded approximately 160,000 species of fly, and believe there is at least this number again yet to be discovered, if not more. Canadian dipterist Steve Marshall believes there may be between 400,000 and 800,000 species of fly in existence, whilst other research points to there being in the millions.

The earliest records of flies are fossils from the Early Middle Triassic period (*c.* 260 million years ago). Once flies arrived on the scene they underwent three main bursts or periods of exciting evolutionary change. The most primitive flies, those first in the evolutionary history, are the Nematocera whose etymology is derived from the Greek words meaning thread and horn in reference to their lovely, long antennae. And not only are the antennae long, they generally have long and slender bodies and legs too. Then around 220 million years ago the first period of rapid evolutionary change occurred and loads of new species of Nematocera suddenly appeared on the scene. After an initial period of new species forming, flies underwent their second dramatic phase of evolutionary change. Some species got chunkier, much chunkier, and around 200 to 180 million years ago what are termed the Brachyceran flies appeared on the scene. The name Brachycera derives from the Greek words for short and horn, as the antennae of these flies are much shorter than those of the Nematocerans. Flies, in a physical sense, evolved from being delicate creatures to being as hard as nuts!

Then about 65 million years ago a final, rapid burst of speciation occurred in the Brachycerans and along came the

Schizophora, derived from the Greek for split and bearing. The Schizophora acquired their name from the way in which they exited the puparium, the rigid external shell protecting the pupal stage. These flies are unique amongst the flies, and most insects, in that they possess this rigid shell formed from the last larval skin encompassing the pupae (well, almost unique as some gall flies also have puparia). To emerge from the pupa, the Schizophora adults blow their heads up into a massive balloon, like a car airbag, which pops open the puparium along a predetermined line of weakness. This is not an air sac; the balloon is filled with hemolymph which can be withdrawn back into the body. Once deflated, the exit point of the balloon on the fly's head hardens to form a distinctive diagnostic ridge on the front of its head called the ptilinial suture – an important characteristic when it comes to identifying the fly.

A final period of species radiation came within the Schizophora and the instantly recognizable species of flies, including house flies, appeared. The calyptrates or sheathed flies, so named because of the membranous pads on the side of their bodies, had arrived on the scene and went on to dominate the globe. You may not know the name calyphrates but it includes families such as the already mentioned house flies along with bot flies and flesh flies. In fact all the generally large and robust bristly flies are found in this group.

The UK is one of the best studied countries for flies in the world as people have been exploring it for hundreds of years and describing everything that they have come across. And that includes world-famous naturalists such as Charles Darwin and Alfred Russel Wallace, as well as others who were less known for their entomological leanings such as Winston Churchill. Museum collections are littered with specimens caught by Colonel such and such and the Right Reverend so and so. In the UK there are more than 32,000 species of terrestrial, freshwater and marine invertebrates of which more than 7,000 are flies, and this means that in the UK one in every five species of invertebrate is a fly. To put that figure in perspective, there are approximately only 5,400 species of mammals on the entire planet. We worry about our lack of knowledge of mammals and about their conservation status, however we also have very limited knowledge of the larval stages of the described species of flies in the UK! Most larvae are very hard to identify, often with very few identifiable morphological structures, and so are just ignored by many biologists. Why should we worry about whether or not we know which larval stage belongs to which adult? Well, with flies it's the larvae that feast, whether on vegetation or on other animals, and so this part of the life cycle can tell us more about the environment than the adults.

So what exactly is a fly? Flies are insects that belong to a group of animals known as diptera, and include not only the ubiquitous houseflies and bluebottles but also daddy long legs, midges and mosquitoes. The name is derived from the Greek words *di,* meaning two, and *ptera,* meaning wings. Indeed all the diptera have two wings, whereas other flying insects such as butterflies, bees and dragonflies have four.

These wonderful creatures go through three distinct stages after the egg: larval, pupal and adult. The immature stages, especially the larval stage, are the ones most affected by any changes in the environment – being less mobile than their parents – and so we can use them to monitor what is happening in the environment. The pupal stage is a truly amazing phase of an insect's life cycle. It is during this part of their development that they reorganize their larval structure to form completely different adult structures, such as wings and reproductive organs. That's worth thinking about – they completely restructure themselves, a process that only happens within some of the insect orders. Once they've emerged from their pupal cases many adult flies don't feed but instead use this time to find a mate and ensure any offspring are laid in the most advantageous surroundings.

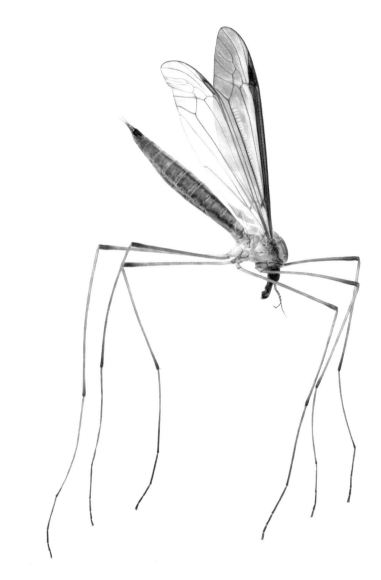

Crane flies, with their slender, long, gangly legs are easily spotted by predators but have the ability to lose their legs and still live.

Once grown, what a wonderful group these various flies make. One of my favourites , the crane fly, or daddy long legs, is familiar to us all and one of the most delicate. The adults, with their slender, gangly legs, can often be seen resting on tussocks of grass, hoping to spy a member of the opposite sex. The long legs enable them to cope with the flexing nature of the long blades of grasses which bend this way and that in even the slightest breeze. Flies resting on these tussocks are rather obvious to other animals such as birds that will try to feed on them. To survive these aerial attacks, the crane flies have evolved the ability to lose their legs when a predator grabs onto them. The bird gets a mouthful of legs while the crane fly is able to escape with its life. The legs, unlike those of some species of animal, do not grow back though so it's not an entirely happy ending, more a spectacular effort to live on and reproduce.

A couple of years ago, in my capacity as curator of flies at the Natural History Museum, London, we reorganised the entire crane fly collection in the Natural History Museum. This comprised looking through more than 300 drawers of specimens collected over hundreds of years from all corners of the world. Over time these pinned specimens had been handled by hundreds of researchers, and flies had been posted off round the world and back for other scientists to study. Underneath the pinned specimens, lining the drawers, we found a graveyard of the fallen limbs of these delicate specimens. So we now have a lovely collection of discarded legs just waiting for someone to develop a key to identify species from leg characteristics alone. We may be waiting a long time.

There is one story from the collection I want to share in this introduction to my beloved subject. It shows the passion that scientists hold for flies, the passion to find evidence, build knowledge and generate innovative thought, but where overenthusiasm needs to be kept in check. This was such an intriguing story that when the truth was uncovered, albeit by accident, it made the national papers.

Even the British tabloid *The Sun* took time off from celebrity gossip to write about it.

The story begins with a fly first described by Johan Fabricius, an important Danish entomologist who specialized in naming insects. From 1792 to 1799 Fabricius published many volumes of *Entomologia Systematica Emendata et Aucta* – great tomes of insect names and descriptions. The volume dated 1794 included a description of a fly described as *Fannia scalaris*, more usually known as the latrine fly. Can you guess where it spends most of its time? I have to say I don't find the genus name Fannia particularly attractive.

Skip to a fly preserved in amber, donated to the Natural History Museum by a distinguished German scientist called Friedrich Hermann Loew. Something of his rather controlled nature can be judged from his refusal to eat warm food while he paid off the loans incurred during his education, and from his extraordinarily precise, machine-like calligraphy. There is never any difficulty in reading a Loew label, with its characteristically neat script taking up the full width. Along with this fly-in-amber specimen, Loew donated his entire Baltic amber collection to the Museum in the mid 1800s. Baltic amber dates from 44 million years ago and this specimen was initially dated by Loew at around 38 million years old.

In 1922, with the amber specimen now in the hands of the Natural History Museum, the fly was critically examined again by German biologist Willi Hennig, the founder of cladistics, the science dealing with the phylogenetic (evolutionary) relationships between species (put more simply, the similarity of one species to another). When he confirmed the amber specimen as the relatively modern-day species *Fannia scalaris* he was surprised. If this were true it was evidence that this species of fly had remained unchanged since this specimen was trapped in amber, suggested by Loew to have happened 38 million years earlier. Hennig published his findings and soon the fly became known as an example of a static

The Piltdown fly, *Fannia scalaris*, preserved in amber at the centre of a 100-year-old scientific hoax.

species, one that hasn't evolved and therefore one that does not fit in with the theory of evolution posed by Darwin just a few decades previously.

It was a revelation. Or was it? Roll forward to 1993. A young curator at the Museum was examining some fly specimens using a rather unreliable piece of lighting kit for his microscope, in that the light source was overheating. What could have resulted in one of the Museum's 'treasures' being irreparably damaged instead led to the discovery of a major piece of fraud. Dr Andy Ross, a palaeo-entomologist, saw (and presumably slightly panicked) that the amber had cracked along a fissure due to the overheating of his light. On closer inspection he realized that this couldn't be the case; that amber was incredibly strong and so wouldn't break so easily despite any dodgy modern-day wiring. A little more detective work and Ross found that the crack had in fact been made by hand, splitting the amber in two. What's more, a small amount of amber had been carved out to make room for, you guessed it, a fly. The forger had then sealed the crack and presented it as a rather spectacular find. Simple. Clever. And fooled everyone for years!

Such is the appetite of the scientific community to explore the natural world, and it is a wonder that this specimen had them fooled for such a long time. They were more prepared to believe in an incredulously old specimen, with little corroborating evidence, than the fact that someone would deliberately try and deceive us. We still don't know who did it or what motivated them but there is a final comment to make. Hennig's paper was published on 1 April – make of that what you will.

What this story tells me is that there is so much we still don't know, that we must continue to study, that we have a responsibility to interrogate, to appreciate and to consider every piece of evidence that comes our way in order to understand the history of our planet and therefore stand any chance of securing its future. Now can you see why focussing our attention on those monkeys and ignoring the glorious fly would be an irresponsible thing to do?

So join me in this book, where with pleasure I'll introduce you into the hidden and ignored wonder that is the fly, from species that throw their eggs while in flight and larvae that completely change in form and lifestyle mid-way through this stage, to the pollinators who live in high altitudes where bees don't go, and so are essential to the floral community, and to the predators that can catch dragonflies in mid-air, suck their insides out and then drop the husks. My hope is that this book makes you look at flies in a different way and think twice about swatting them.

My work at the Natural History Museum, London has enabled me to play (really that should be work but I would be lying) with one of the best collections of flies in the world, meet some of the most interesting and often eccentric people who study them, and travel to the far corners of the world collecting and observing them. I have a lot to learn about these intriguing creatures but then I am not alone in that. This is not a book on the taxonomic arrangement of flies (that is, their relationship to the rest of the animal kingdom), or on their structure or workings. Rather it is a book of my ramblings through the world of flies and how they interact with everything else in the environment, sometimes to the detriment of humans but more often than not to our benefit.

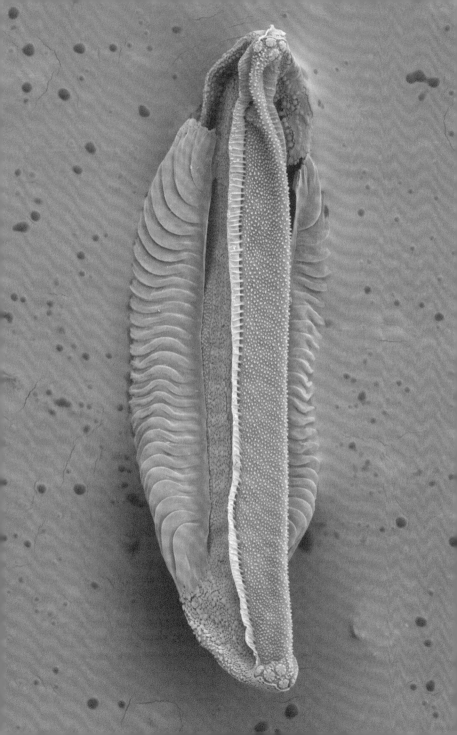

CHAPTER 1

The immature ones

Listen up, maggots. You are not special. You are not
a beautiful or unique snowflake. You're the same
decaying organic matter as everything else.

Tyler Durden, Fight Club

A FRIEND OF MINE once called me about a specimen she wanted
me to examine. She was a vet and had been contacted by
a police dog handler about a problem. I am used to getting insect
enquiries and so this was not, as such, an unusual request. One of the
police dogs had, to put it politely, excreted a rather large maggot from
its derrière and she hoped that I could determine what it was and
whether it was harmful or not to the dog. To be truthful, I was not
expecting a maggot at all as most of the time the things that come out
of a dog's backside are parasitic worms or helminths. But being a vet
she would have picked up on this so I was a little curious. I was even
more intrigued when I opened the package and there indeed was a
maggot, but not the type I was expecting. There are many larvae that
specialize in living in other animals, but this one was just a free-living
larva called a leatherjacket, a common name for many of the crane fly
larvae. These are long, cylindrical larvae, rather nondescript in form

A scanning electron micrograph image of the egg of *Anopheles arabiensis*,
highlights its rippled float and its breathing florets.

except for their most exquisite anal breathing tubules! This was very odd, I thought. Not just because it had emerged from an unexpected place but also because it was completely intact. How had the specimen managed to come through the digestive tract apparently unscathed? The ability of maggots to survive in the most hostile of environments is amazing and well recorded, but in all probability this one could not have gone through the entire length of the dog's digestive tract, which is much more acidic than our own, and remained in such good shape (albeit dead). It was more likely that it had entered via the rear whilst the dog had been sitting on grass.

This may seem a random anecdote but the point I'm making is that maggots can turn up anywhere, and their ability to survive in some very odd places is one of the reasons why flies have become so globally successful. We often focus too much on the adult stage, which can last as little as a few hours. The maggot or larval stage of the life cycle is often the longest and most indulgent stage of a fly's lifetime. The lovely larvae can spend years just devoting themselves to eating and eating, and why forever not!

As discussed in the Introduction, the life cycle of a fly is divided into four stages: egg, larva (or maggot), pupa and then adult. Flies go through a complete metamorphosis, and the larval and adult stages are completely different in form. The larval stage is all about eating, and by gum they do it well, whether their food is dung, rotten flesh or a juicy carrot. The adult stage is often just about sex and dispersal.

Every maggot begins as an egg, externally deposited by a simple egg dispersal structure called an ovipositor, an egg-laying tube. In some species this structure has been modified into a very impressive tool that can penetrate a host's body and lay eggs inside it. Dipterists will often go into raptures when discussing the male genitalia of flies as they tend to be large and complicated in structure, often supporting some very odd appendages. You will also hear disappointed mutterings on diptera-collecting trips when females are 'pooted up' (sucked up into tubes) as it is very hard to distinguish between females of a species due to their simple genitalia. But females blessed with ornate ovipositors need not feel inadequate because their appendages can be equally large and complex. The general public are often alarmed by some of these structures because they are very similar in appearance to the stings of bees and wasps. In fact a sting is a modified ovipositor but, unlike with bees and wasps, flies don't use their oviposters to deliver any venom.

As far as diptera are concerned, ovipositors, when present, are generally tapered towards a straw-like end, as seen in fruit flies and their relatives. As already stated, they are not stinging weapons and I have yet to encounter or hear of anyone being attacked by a fly's rear end – attacked by the head-end all the time but not by their rear. Don't get me wrong – to other creatures and even ourselves some flies do herald death from this end by dispersing larvae which eat their hosts from the inside as they develop. Many species of fly have parasital or parasitoidal larvae. Parasital ones leave the host damaged but rarely is this terminal, whilst parasitoidal larvae are nearly always fatal to their host. With the latter the flies either lay their eggs internally or on the surface of their victim. In one group of parasitoid flies, the Conopidae or thick-headed flies, the female has what is often described as a clamp on the end of her abdomen where she grabs onto the host and slices it open as you would a tin of anchovies. Tachinidae are another family where all members are parasitic and so these are commonly all called the parasitic flies. Females also often lay eggs onto the hosts' food plant so they will be ingested, or they scatter their eggs over a wider area or, with great parental care, glue them onto the body of the host. The hosts may walk around for several days with the eggs glued onto them before the larvae develop enough to be able to penetrate the body.

The egg stage is exactly that, an egg which may be laid as a single unit or in clusters, with numbers reaching into the hundreds depending on the species of fly. Eggs of the genera *Tabanus* and *Hybomitra*, in the horse fly family Tabanidae, are laid in multiple layers of around 500 eggs. They look something like a chocolate-coloured Jenga and can be found on the surface of streams and on branches over wet ground. One of the reasons why the common fruit fly is such an excellent species to study is because they are so fecund (although not quite as fecund as the horse flies as the females only lay around 400 eggs) and this, coupled with their short generation time, enables us to observe genetic change rapidly.

Many parasitoid flies produce vast quantities of eggs in one batch and scatter them around on the surrounding vegetation to maximize the chance of a larva hatching near a host. Some of these eggs will remain in situ until they are munched up by an unsuspecting caterpillar. Once the caterpillar has inadvertently ingested the egg, development is triggered and, within the host, a larva hatches and

The eggs of *Tabanus* sp. of the horse fly family, are neatly laid in multiple layers of around 500 eggs.

starts eating the caterpillar's insides. Living in another organism is a wonderful environment for the little ones to develop in – it is warm, safe and there is plenty of fresh food within reach.

Other species lay their eggs in or near water or in damp, moist environments. This variability in habitats has led to huge variety in the shape, structure and numbers of eggs deposited. Many are highly sculpted with pits, ridges and craters across the surface. These indentations enable a pocket of air to remain next to the egg's surface, which in turn allows oxygen to diffuse into the egg and prevents the organism from drowning. Luckily for us researchers, these indentations can also be very species specific and so are helpful for identification purposes.

Flies in the mosquito family Culicidae lay their eggs in many different aquatic habitats, from grazing marshes to salt pans, from coastal pools to tree holes, and they often take advantage of temporary environments such as puddles. Recently humans have unintentionally given them a helping hand by providing ideal oviposition sites in the form of vast mountains of tyres. Most female flies opt for the lay-and-leave strategy and mosquitoes are no exception, but the number of eggs and how the eggs are distributed does vary across the family. Mosquitoes are divided into two subgroups: the Anophelines and the Culicines. The former includes the females that transmit, amongst other things, malaria. The latter don't pull their punches either when it comes to disease transmission, being willing and able vectors of diseases such as dengue and yellow fever.

The females of the mosquito genus *Anopheles* lay between 50 and 200 single eggs at a time. To stop these eggs from sinking, and subsequently suffocating, they provide built-in floats. Imagine an oversized hotdog floating in a bun and you have the correct image in your head. These eggs are incredibly sculptured and have several breathing tubules resembling broccoli florets at each end.

Culicines have adopted a different strategy of egg-laying. In the genus *Culex*, females deposit their eggs in clusters and, instead of laying them flat on the water's surface, they lay them together in vertical lines, so forming a buoyant structure that resembles hundreds of little bowling pins glued together. One cluster or raft may contain up to 300 eggs. Although it is the most common practice, not all Culicine eggs are laid on the water surface. One genus, *Mansonia*, has taken to gluing its egg masses to the underside of vegetation or on the roots of trees and palms in swamp forests.

The time that it takes an egg to develop depends on the species, and some can even delay their development by entering a dormant stage if conditions are too adverse, such as during winter. Other flies completely remove the risks associated with laying their eggs in the external environment by letting them develop inside them. When your average female crane fly exits her pupal case, she will already be full of mature eggs ready to be fertilized. Males are never far away and she usually does not have to search for long before she finds a male to copulate with and then quickly lays her eggs. I say lay, but

The eggs of the *Anopheles costai* mosquito are laid in aquatic habitats; to stop them sinking they are nestled in a float.

The mosquito *Mansonia* sp. glues clusters of up to 300 eggs to the underside of vegetation.

they have been observed just dropping their eggs in mid-air. Bee flies in the family Bombyliidae are even more spectacular in their egg dispersal. These aren't just dropped mid-air but flung into or near the nest entrances of solitary bees, wasps, beetle larvae, grasshoppers and ants, so the larvae can develop on these hosts.

The egg stage is the shortest one, with some larvae hatching out after just a few hours. And some species of flies forgo an external egg stage entirely. There is a rather attractively named group of flies called flesh flies in the family Sarcophagidae, which feed on all sorts of nutritious substances such as dung, carrion, decaying flesh and open wounds – the name is most apt. These flies depend on material that by its very nature is short-lived. I undertook an investigation recently with a colleague to record the species of flies living within cow and horse pats and to understand more about the feeding habitats of these specific species. These pats completely broke down in the lab after a couple of months through microbial as well as invertebrate activity. This would have happened at a faster pace in nature because of better environmental conditions and increased competition from other dung lovers rushing to get to the limited resource. For animals that utilise such a temporal food source any advantage given to the developing offspring is obviously a huge help. Flesh flies do this by giving birth to live young, a term referred to as viviparity. This ability has evolved not just in flesh flies but in many other species, especially within the calyptrates, and often occurs with larvae that go on to have less than palatable diets in terms of our sensibilities – be it rotting meat or stagnant faeces. Giving birth to live young rather than eggs is a good thing for all as it means that the larvae are immediately ready to start consuming this temporary habitat of decomposing bodies and faeces.

How the young that develop inside the mother are fed is fascinating. There are four or five closely associated families (within a superfamily grouping called Hippoboscoidae) where all of the adults are bloodfeeders and all of the larvae develop inside their mothers. The larvae are all nourished via an internal milk gland that produces two milk proteins and endosymbiotic (internally mutualistic) bacteria. This method of advanced parental care is termed adenotrophic (gland-fed) viviparity. Eggs are produced one at a time and retained internally, and they contain enough yolk for them to develop into larvae.

One of my favourite families out of this group is the Nycteribids – the bat lice flies. They are very odd-looking flies indeed, with the wingless adults looking more like skinny spiders that lost a pair of legs in a brawl. As their name suggests, they feed on bats as adults. The adults cling to the bats most of the time, except when a pregnant female crawls off the bat and deposits her fully grown larvae onto the cave wall. Upon leaving their mothers the larvae almost immediately pupate into a dome-shaped puparium. The mother then reverses over it with, squashing it with her backside to ensure it is properly secured against the cave wall.

Giving birth to live young is not restricted to Hippoboscoidae; a total of 22 families bypass the egg stage. Weirdly it is more common for larvae to be internally brooded as singletons (unlarviparity) or in large numbers (multiviviparity) than in intermediate-size groups of two to 12 eggs (oligoparity). *Ocydromia glabricula,* in the Hypotidae family, is one such species that practices multiviviparity, and it has been observed flying over faeces and dropping multiples of larvae onto the dung below.

The number of viviparous flies is small in comparison to the number of flies that lay eggs though. The common house fly lays up to 500 eggs in her lifetime, but this is nothing in comparison to some parasitic flies that can produce thousands of eggs, though never all in one go! The fact that we are not overrun by flies is testament to the number that perish along the way due to factors such as predation and desiccation.

Once hatched, the larvae differ greatly in form across the order, especially between the primitive families of flies, the Nematocerans, and the more advanced families, the Brachycerans. Primitive families have a more defined form and are termed eucephalic due to their distinct head capsule with chewing mouthparts. These are also culiciform (gnat) shaped. Brachyceran larvae have reduced or no head capsules – their mouthparts have been reduced to hooks – and they are called vermiform (literally meaning worm-shaped).

The vermiform species split further into two groups: the hemicephalic larvae, which have an incomplete head capsule and partly retractable mandibles, and the acephalic species, which have no distinct capsule and mouthparts forming a cephalopharyngeal skeleton. Yes, a cephalopharyngeal skeleton! That is, an internal skeletal structure that acts in a similar way to the human jaw, located in the larvae's head region. And it is this latter group that we colloquially refer to as the maggots.

The larval stage is the most difficult to identify as a species as it often lacks distinct morphological features. Species with complete head capsules may be easy to identify as there are distinct features such as teeth or spines, but the more advanced species, with their not-so-easily pronounceable cephalopharyngeal skeleton, resemble nothing more than a mini, see-through sleeping bag with a mangled clothes hanger sticking out of one end.

It's not all doom and gloom for taxonomists however as there are other features apart from the head that are useful for identification. For example, larvae have breathing tubes, known as spiracles, which form part of their breathing system and vary across the fly group, with there being seven different ways in which they are set up. The most common arrangement is called amphinuestic (from the pneustikos meaning to breathe), which is when the larvae just have one pair of spiracles located near their head and another pair, more prominent in some species, near the apex of the abdomen. Other arrangements, such as the proneustic systems only found in some fungus gnats, are more specialized, whilst others have taken their basic spiracle form and run with it (as it were). The larvae of some hover flies, the drone flies, are called rat-tailed maggots and with good reason. Their posterior spiracle is very long and enables them to reach the open air even when the rest of the maggot is feeding in some pretty dismal habitats such as manure pits. And the cranefly larvae have the most fanciful arrangements of them all, with their anal spiracles making them look like miniature, hairy monsters.

Returning to the mosquitoes, the Culicine larvae all have an elaborate structure – a very large breathing tube called a siphon – arising from their abdomen . The larvae dangle down through the water column with their siphon protruding through the water surface, thus enabling them to breathe air directly from the atmosphere. This is essential as these species, like the rat-tailed maggot, are often found in stagnant habitats with very little available

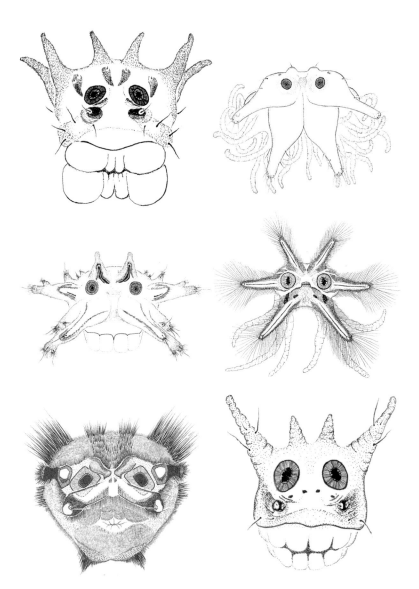

Crazy Tamagotchi? The spiracles or breathing tubes of crane fly larvae look like they were created to resemble evil Tamagotchi characters.

oxygen in the water. If you disturb the surface they all shoot down to the bottom to escape – I could, and have, spent hours poking into pools with mosquitoes in to see this escape behaviour. Their motion is purposefully erratic to avoid predation and they are really fast!

Anopheline larvae, from the other mosquito subfamily, have no such breathing tube and so have to rely on smaller spiracles located at the end of their abdomen. For them to obtain enough oxygen, the larvae lie horizontal to the surface to ensure maximum contact. And that is the simplest way to distinguish between the two subfamilies at this stage – dangle down or lie flat, siphon or no siphon, respectively. Interestingly they also exhibit different resting stages as adults but the other way round, with the culicines lying flat and the anophelines conducting what look like handstands with their back legs up in the air.

The next stage of a mosquito's life history, as with all flies, is the pupal stage, and with mozzies these are affectionately referred to as tumblers. Mosquitoes differ from most other flies here in that for them this stage is an active one as they wriggle around inside their pupae. The pupae often position themselves just below the surface, with their breathing tubes – their trumpets – reaching through the surface membrane to the air above.

James Barbut, an English naturalist and painter from the eighteenth century, produced a book of illustrations of the species Carl Linnaeus had previously described in *Systema Naturae*. Barbut not only included diagrams that, although not necessarily completely accurate, are wonderful to look at, he also wrote passionate descriptions and tales about the species he drew. His description of adult mosquitoes emerging from the pupal case is pure poetry: 'The robe he lately wore [the pupal skin] turns to a ship, of which the insect is the mast and sail'. Watching them emerge is indeed a glorious event, but it is also a perilous stage and many things can and do go wrong. I have reared many mosquitoes from birth both at home and at work – I once even had to go into work on Boxing Day to look after the 'babies' – and have seen a fair few adults die when trying to exit from the pupal case as they get stuck in their skins. Females often have an eventful and sometimes perilous emergence as males tend to emerge first and lie in wait for them, pouncing before the females have completely shed their underwater skins.

Until recently mosquitoes were classified into three subfamilies – the additional one was called Toxorhynchitinae of which there was a single genus *Toxorhynchites*, but this is now placed within the Culicines. *Toxorhynchites* have very different life histories in relation to other mosquitoes in a way I think most humans would approve of – both the male and female adults are vegetarian, so have no appetite for human blood. Thus no incessant whining around your head at night whilst they locate their targets, no nuisance biting resulting in irritating lumps that take days to disappear. Not only that, their larvae are predators of other mosquitoes! *Toxorhynchites* are big mozzies at both the adult and the larval stages. If you happen to catch some larvae in your sample along with other species, you need to separate these monsters out immediately as I learnt the hard way in South Africa a few years back. I was collecting mosquitoes from some wonderful wetland habitats and some not so salubrious ones including a disused and stagnant swimming pool. I collected many larvae of all species to rear into adults but was remiss in separating them out that night. Lo and behold the next morning there were fewer, albeit now larger, individuals. The *Toxorhynchites* among them had been feeding on the others.

The ability of these predacious larvae to consume large numbers of other mosquito species led to them being investigated with a view to determining their efficacy as biological control agents. Although in its infancy, the use of *Toxorhynchites* to control the more dangerous mosquito species is a really appealing idea as it limits the need for more expensive and sometimes toxic chemicals or other types of biological control which are not so selective in diet, eating both the harmful and beneficial species.

The problem can be getting these advantageous larvae into the areas where the other nasty species (in terms of disease transmitters) reside. This is relatively easy with species where the larvae live in open habitats such as pools, pits, ditches and ponds, but there are other species of mosquitoes that live in more cryptic environments such as tree holes and old tyres. There is even a species of mosquito living underneath the streets of London, in the Underground. This sub-species has been causing problems for a while and was a notable biter of the poor residents of the Underground during the bombing of London in World War II. *Culex pipiens f. molestus* is in the species group *Culex pipiens,* and *molestus* and *pipiens* are so similar we were having difficulty separating them morphologically. The 'f' represents 'form' and indicates that there is a behavioural difference between the two species that may or may not be enough to cause the species to separate. The *molestus* form is only found underground but is not restricted just to the London Underground – it has been found underground all over the world in both natural and human-made systems. We don't know why the larvae live underground and really don't know much about this form except that the female is a vicious biter. The stories of people bitten by the adults during the Blitz sound horrendous though maybe not when compared to the experiences of the majority of people who were sleeping above ground.

It may be difficult to believe that, apart from some biological control agents, there are any benefits to be derived from mosquitoes (apart from their aesthetic beauty). However, the large densities in both the larval and adult stages provide an invaluable food source for both birds and fish. Mosquitoes evolved over 79 million years ago and have spent this time adapting to their environment and, in turn, their predators have adapted to feed on them. If you choose to wipe them out you may subsequently wipe out other, as the press puts it, more 'charismatic' species, the megafauna. All of those majestic migratory birds or freshwater fish, important in terms of biodiversity or food security, would die out. Trout for instance have a diet rich in insect larvae, including many, many mosquitoes. Wiping out a species from the environment usually has unintended consequences, doing more damage than it is initially thought it would do.

One fascinating adaptation mosquito larvae have developed to survive in inhospitable environments is the separation of their breathing and feeding components. By having elongate spiracles, those on the body as opposed to around the mouth, they have freed up the mouth from having to play any part in respiration and so it can solely concentrate on ingesting food – just imagine 24/7 eating. The larval stage of any insect is the one generally designed for eating, and if life for the rest of us is 'be born, eat, shag, die', for most insect larvae it is much more like 'be born, eat, eat, eat, eat, eat, shag, die.' As fly larvae don't reproduce or fly or even move much, they have

An example of a net-winged midge larva from the Blephariceridae family, showing the prolegs along the body.

a much simpler body plan – they do not have true legs and are affectionately (by me at least) called the legless larvae. Some have false legs; some non-biting midges have one pair of fleshy feet at the front and back, or like the net-winged midges, the Blephariceridae, have prolegs down the body – but neither of these two examples are true legs. This general lack of the limbs that other insect larvae possess is because they are highly specialized examples of precocious larvae – i.e. very early hatchers from the egg. And this is what arguably enables flies to exploit the most diverse range of habitats of all insects. Their elasticity means that they can squeeze themselves between surfaces and into tiny holes and therefore take advantage of so many different food sources.

Of all the fly larvae adaptations, one of the most astonishing must be their ability to live in salt water. Insects were not thought able to survive in this habitat but we now know they do. There are 12 genera including *Clunio, Thalassomyia* and *Pontomyia* that have marine species. We really don't know much about them; in the *Pontomyia* genus only four species have been described and one of these just from the skins of the larvae and pupae. To describe a species from either their larval or pupal skins is not unusual, especially with midges and mosquitoes, but to avoid confusion we try not to do it. Often, years later, we discover that we already knew the associated adults but had called them something completely different.

A close competitior, in terms of most astonishing adaptations, is the petroleum fly, *Helaeomyia petrolei,* from the Ephydridae or shore fly family. This species is in fact found a long way from the shore, preferring instead to live in the extreme environment of petroleum pools in California, USA. The larvae ingest oil and asphalt, resulting in a lovely petroleum-filled stomach. They are not actually feeding on this mix but rather on the particulate matter that has fallen into it. This is one of the more extreme fly habitats, with not only a difficult and dangerous substrate but also one that is very hot – they are dealing with temperatures of up to 38°C (100°F). It is not the most extreme however. Research has found that a dehydrated larva of the species *Polypedilum vanderplanki* can survive in liquid helium – a temperature of -270° C (-454° F) – for five minutes and still live.

As with some eggs deposited as a cluster, or multiple larvae developing within the mother, some of the free-living larvae have developed a communal lifestyle. I was once brought a most exciting specimen of the bracket fungus *Ganoderma applanatum* by one of the botanists at the Museum. It was not the fungus that intrigued me but what was on it, hundreds of abnormal growths called galls caused by the larvae of *Agathomyia wankowiczii* (I guess when it was named more than 100 years ago that didn't sound as silly as it does now). This is a type of fly called a flat-footed fly from the family Platypezidae – aptly named as they have enlarged hind tarsi or back 'feet'. This relatively small family is commonly called the yellow flat-footed fly in the UK, and there are about 33 species in the country. The one in question has only recently arrived in the UK and it's one

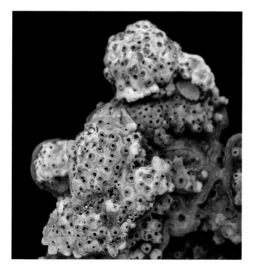

Galls on bracket fungus, *Ganoderma applanatum*, created by the larvae of the yellow flat-footed fly, *Agathomyia wankowiczii*, one of only three insect species to form galls in fungus in Britain.

of a select group of gall-forming insect species (there are two further species) in fungus in the UK – it's a very special fly. Most species of gall-forming insects can't grow in fungus as they decompose too quickly, but this is a slow-growing, frugal species which forms hard brackets that do not decay before the grubs have developed. Once fully grown, the larvae bore a hole at the top of the gall, which is technically the bottom as they are on the underside. Gravity does the next bit and once they have landed on the soil they dig down into it before eventually pupating and emerging as adults.

Fly larvae don't just get everywhere, they also do extraordinary things. My mother travels a fair bit as she is a woman of leisure these days, and annoyingly for me has seen things that I am still so very desperate to see. One of these involved going to New Zealand, more specifically the Waitomo Glow-worm Caves on North Island. Glow-worm is a term generally used for beetles, but the larvae of two gnat families of flies also demonstrate the ability to glow. And unlike the harmless little beetles that glow to attract mates, the fly larvae glow to attract their prey, luring them to their death by trapping the attracted prey on threads they dangle down. You have to admire these larvae. They have turned their own bodies into lures to trap their prey.

Orfelia fultoni is another bioluminescent species from the fungus gnat family Mycetophilidae and is found in North America. It uses two bioluminescent lanterns at either end of its body to produce a blue light that attracts other insects from along the stream banks where it resides. In fact *Orfelia fultoni* have the bluest of all light emitted by luminescent insects. The classic *Spider and the Fly* poem by Mary Hewitt is reversed in these cases as the glowing fly larvae dangle threads beneath them to trap their prey, which have been attracted by the lights of the insects above them. *Oreflia fultoni* also spin webs and lie in them in an exceptionally similar way to orb spiders. However their webs have more the style of Picasso about them rather than the precision of Leonardo de Vinci.

So the larval stage is an exciting, action-packed one in a fly's life cycle, and larvae show an extraordinary zeal for life. Fungus gnat larvae, for instance, feast for between six and 12 months before pupating and turning into an adult, the timing depending on the quantity and quality of their food. And I've always had a fascination for the botfly larvae's determination to find flesh to eat. Known as *Cordylobia anthropophaga* – anthropophaga means human eater – the tiny flies lay their eggs on, amongst other things, damp clothes and linen. A colleague who lived for a while in East Africa told me how everyone was warned to iron everything after it had been hung on the clothes line for fear of these larvae, and so her mother was always very careful to do this. However, there was one lapse involving a school skirt with an elasticated waist. The mother carefully ironed the skirt but did not stretch out the elastic and so, unbeknown to her, the eggs that were laid survived and the larvae hatched next to my colleague's skin and burrowed into her belly. These, as with many parasitic fly larvae, burrow under the skin of many mammals including humans and stay there until they drop out to pupate on the ground. They are a serious problem in tropical Africa and they are the commonest form of myasis, a debilitating parasitic infection caused by flies. My colleague holds no grudge against these little creatures, in fact she works with them along with many other maggots. Her mother is still mortified.

As Harold Oldroyd wrote in his wonderful book *The Natural History of Flies*, the larvae and adults of flies are more different from each other than those of many other orders of insects. And, in many ways you could argue that they fit two lifetimes into one as they differ not just in form between the larvae and the adult but also in how they live. Larvae may be the immature ones, but they are a source of endless fascination.

CHAPTER 2

The pollinators

What kind of monster could possibly hate chocolate?

Cassandra Clare - Clockwork Angel

Hate chocolate? Well, I do. I simply detest the stuff – have done for years. I dislike the texture and the way it slimes down your throat, but most of all I don't like the smell – just thinking about it turns my stomach. Even I have to admit this is not the most normal of dislikes. It is ironic considering my love of flies. Confused? Flies, you see, are the only pollinators of chocolate, or more specifically *Theobroma cacao*, the cacao or cocoa tree. This plant species has a complex reproductive structure, so complex in fact that only one group of very small flies, amusingly known as No See Ums, can pollinate it. This group, from the *Forcipomyia* genus of the family Ceratopogonidae, are, along with the rest of the family, known as the biting midges. Biting midges are cursed across the globe for ruining many a day in the countryside, especially the infamous Highland midge swarms in Scotland. According to her diary, Queen Victoria was half-devoured by these little ladies whilst at a picnic in Sutherland woodland in 1872.

The hirsute form of the male chocolate midge, *Forcipomyia* sp., which is essential for cocoa pollination. The pin obscures one of the legs.

The female adults of many biting midges have very painful and sometimes fatal bites due to some of the diseases they can transmit. They can also swarm in huge densities which can cause a large amount of blood loss.

But without these minute, often rage-inducing flies, many people would consider that life is no longer worth living. Cacao producers are very worried about the ongoing supply of their 'miracle' substance. Our (yours not mine) demand for the stuff is vast – it's an $80 billion year industry with 3.5 million tonnes produced annually, a figure set to increase to 4.5 million tonnes by 2020. But it is a volatile time for this product – traditionally the tree has been grown on small-scale farms but these are affected by increased stochastic weather patterns, growth in the numbers of pests and diseases and by political instability in many of the countries where the cacao tree is grown. These factors are exacerbating the already naturally low pollination rate of the plant. Many of the small-scale cacao farms are now moving across to larger set ups to overcome some of the negatives and boost yields, but this has also had repercussions on production.

The cacao plant has both male and female reproductive organs on the same plant but it cannot self-fertilize and so is entirely dependent on the midge to do this. It's one of the many little known facts about flies and the benefits they bring us. Despite the efforts of the flies, few of the flowers go on to produce fruit; it's a tricky business. No fruit means no bars of Galaxy or Curly Wurlies. Add poor pollination rates to the cultivation of these plants and your success rates drop further. The fact is, the pollinating flies are tree lovers – they like damp and shady conditions and many of the species require aquatic, semi-aquatic or moist soil conditions for their larvae to develop in. On these cultivated farms, trees are removed to create more space for the cacao but this removes most of the shade as well. So now there is very little shade and very limited leaf litter, and so nowhere for either the adult flies or their offspring to live.

In cultivated plots the average pollination rate is shockingly low at 0.3%. By cultivating forests for the production of the plant we are ironically destroying chocolate (and the midge!). Surely for chocoholics saving this fly ranks up there with saving the giant panda, which also has reproduction challenges?

Not only do flies from this family pollinate chocolate, some – again in the genus *Forcipomyia* – pollinate heather species within the genus *Erica* (I was destined to love flies). Why is this important? Well heather grows with abundance in the Scottish mountains, and the Picts – a group of people who lived in what is today eastern and northern Scotland during the Late Iron Age and Early Medieval periods – used to make heather ale as long ago as BC 325, and that they could was largely thanks to flies. The myth behind the birth of whisky was that some of this heather ale was being baled in a stone roof cottage, the steam condensed and into a cup, whisky dripped!

That story may not be totally true but there are many similar stories that reveal the close relationship between insects and flowering plants. Flowering plants evolved after insects had arrived on the planet and the success of plants is very much down to the symbiotic relationship they have with insects. Evidence of this relationship first appeared between 130 and 140 million years ago and the intervening years have seen many plants and insects, including the flies, evolve some exclusive relationships with each other. Two species of biting midge have been observed pollinating the long flowers of *Erica* species and are only able to do so as they have elongated mouthparts – so the nectar in these plants is only accessible to insects with long proboscises. Another fly, *Rhynchoheterotricha stuckenbergae*, a small dark-winged fungus gnat from the family Sciaridae with a true tongue-twister of a name, is another species with a very long proboscis, about three times the length of its head! *Peringueyomyina barnardi,* a primitive crane fly in the family Tanyderidae, has an equally long proboscis and another crazily long name – maybe there is a correlation between the length of names and the length of the proboscis? All of these species have only been found in the Cape region of South Africa where the climate and geography have led to the development of a highly endemic flora which includes many species with nectaries found deep within the long, tubular petals of the plants. These long-named flies and long-tubed plants have become co-dependent on each other.

The pollinating role of flies is hugely important for the general health of a range of ecosystems, including agricultural ones. Of the 150 families of flies, almost half, 71, have been shown to feed from flowers and therefore in principle transmit pollen from one plant to another. It's not just the number of species that qualifies flies as important pollinators but also their distribution. As already mentioned flies are ubiquitous – they are everywhere. My colleagues and I have caught flies in the most unlikely places, the toughest of which was at more than 4,800 m (15,750 ft) altitude, up a mountain in Peru. This is not easy as the 'pooters' we use to catch them rely on manual suction. I don't know how many of you have been to those altitudes but the oxygen is so thin I could barely crawl, let alone breathe. We had to try and suck up flies into a tube, often with limited success, with many a fly just sitting there on the plants unaffected by our ineffectual pootering. The point of this story is not to highlight my own inadequacies in failing to collect flies but to point out that flies are found at such high altitudes. A whole variety of them are found in these regions, including many hover flies from the family Syrphidae.

Hover flies are exceptionally common, distributed everywhere and very species-rich, with more than 6,000 species described globally to date. They are considered to be the most important of dipteran pollinators although this may change as our knowledge of fly biology increases. As their name suggests, they hover – when I was learning Latin family names I always found this group easy to remember as I would think of them 'surfing' ('Syrphing') on the wind.

The wonderful bee-mimicking hover fly, *Volucella bombylans*, looking like a more dangerous species so as to protect itself against predators.

In the USA this family is also called flower flies in acknowledgement of their associations with plants and their importance as pollinators. Many species in this group are distinctive and familiar in appearance – but not as typical flies. Rather, they are clever mimics of bees, wasps and hornets. It makes sense for a species to look like a more dangerous species to protect itself against potential predators, which at a glance will ignore them for a less risky morsel.

These pollinators have no venom like their dangerous doppelgangers, let alone a sting. But they do spend a lot of time out in the open, guarding their territories and trying to attract the opposite sex or feeding from plants. Most people have no idea that most of the little yellow-and-black insects zipping round their garden at speed – some can fly in short bursts at up to 25 mph (40 kph) – are not the helpful bees they think they are, but rather helpful flies. Among their many morphological adaptations to assist with pollen transfer is their covering of thick hairs which, while the fly is feeding on flowers, pick up pollen.

Hover flies are second only to solitary bees and bumblebees in their value as commercial pollinators. The economic worth of all insect pollinators of cultivated crops has been estimated at about £120 billion, which equates to 35% of global crop-based food production. Flies form a high component of that figure and they are key pollinators for many crops including mango, chilli pepper, black pepper, carrot, fennel and onion. I may not like chocolate, but I would be devastated not to have pepper in my life.

One of the most spectacular-looking groups of pollinating flies is the tangle-veined flies of the family Nemestrinidae. The adults are beautiful creatures, generally robust and chunky in shape and, more often than not, fluffy. As their name suggests, they have a distinct vein pattern on their wings which helps us to identify which flies belong to the family. And with this family, very unusually, it's the female's genitalia that grab the attention of dipterists. Male genitalia in flies are often ornate and vary considerably across species, even closely related ones, whereas with females there are generally few or no differences. Within Nemestrinidae however the female's ovipositor varies considerably across species and so it is a very useful characteristic to aid identification. There are five subfamilies of tangle-veined flies: two of them (Hirmoneurinae and Nemestrininae) have telescope-shaped ovipositors that have retractile segments forming a pump-action, egg-laying machine. The other three subfamilies (Atriadopsinae, Trichopsideinae and Cyclopsideinae) have sabre-shaped ovipositors, with two very long and slender valvulae (a scientific term for diptera lady bits) from which the female shoots her eggs.

It is not just the shape and functionality of their genitalia that make the females so unusual, they can also lay several thousand eggs in their lifetime – compare this to a house fly, which only lays about 500 eggs in a lifetime. Several thousand may seem like a lot of eggs but there is a high attrition rate because the food source sought by the larvae is rather mobile, and includes locusts and grasshoppers. After about 10 days, the scattered eggs hatch into very active larvae called planidia. These readily disperse, often helped by the wind, into their surroundings to seek out their hosts, which vary across the family, with the Nemestrininae and Trichopsideinae subfamilies choosing grasshoppers, Atriadopsinae seeking bush crickets and Hirmoneurinae preferring scarab beetles. The fifth subfamily, Cyclopsideinae, is only known from a couple of specimens found in Australia and we know nothing of their host preference or anything else about them. This subfamily is solely represented by those few specimens and comprises just one species. What is depressing, but sadly a common state of affairs with many species, is that the holotype – the specimen that was used to formally describe the species – was destroyed. Not only do we know nothing about the biology of these flies but we also have hardly anything to work with in museum collections.

HOLOTYPE
Cyclopsidea
hardyi.
Mackerras.

Destroyed while
in Hardy coll.

Australian Museum
K 359371

Holotype of *Cyclopsidea hardyi*, or rather what's left of the holotype
- a taxonomist's heartbreak.

What we do know about the larvae of the other subfamilies is that they can survive for up to two weeks as an active mobile planidium, seeking out a host. Their hosts are also very active and most larvae perish during this stage. Successful individuals then undergo a second morphologyical change once they have found and penetrated their host. Not content with normal larval development, they alter the structure of the later stages hugely in comparison to the initial one, a process known as hypermetamorphosis. After being the sleek, active little host-seekers, they then behave like slobby teenagers by becoming sedentary parasites.

Although the larvae are interesting, it is not this stage that is important in pollination. Adult tangle-veined flies are special pollinators in that many are species specific and have co-evolved with their host plants to such a point that the flies often have the most spectacularly elongated mouthparts, necessary to penetrate the equally long tubes of the flowers in exclusive, mutually dependable, relationships. Most nemestrinids, of which there are approximately 330 species globally, have fairly long proboscises. Many of these are fairly rigid and can't be curled up neatly out of the way when not in use as seen with butterflies and moths. Flies cope with what would seem to be an enormous inconvenience by loosely tucking the proboscis underneath their body as they fly, and a few species are able to partially retract them into their heads.

The idea of manoeuvring around with a proboscis 0.5 cm (¼ in) long for flies around 1 cm (½ in) in length sounds fairly cumbersome, but that is nothing in comparison to one species, *Moegistrorhynchus longirostris,* which has a proboscis that can reach eight times its body length (that's up to 8 cm or 3¼ in). If humans had a tongue equal in ratio to this our tongues would be over six metres (19½ ft) long. This species has the longest proboscis in relation to body size of any insect. And why is it so long? Because they have co-evolved with the plants with long-tubed flowers, which include irises, orchids and geraniums, to feed and pollinate them exclusively. These

Not all fly proboscises can be curled up neatly. That of *Hirmoneura anthracoides* is fairly rigid and it can only tuck the proboscis under its body as it flies.

groups of flowers are either pollinated exclusively (eight species) by *M. longirostris* or by it and a few other morphologically similar species. *Moegistrorhynchus longirostris* is therefore a keystone species, in that it is critical to the survival of these long-tubed plants – remove this and the flies die out. The Cape region of South Africa, where these flies are found, has an internationally recognized flora due to the diversity of plants that exists there and, when we get round to identifying all the fly species from this region, we'll probably find they are equally diverse (and, may I add, equally if not more attractive).

Another dominant group of flies found in this region are horse flies in the family Tabanidae. Horse flies, when they are acknowledged, are generally considered evil, annoyances of impressive size and endeavour. But most of the females of these determined bloodfeeders also need nectar to provide energy for flight and the males feed exclusively on it. Within the subfamily Pangoninae, commonly called the long-tongued horse flies, there are many examples of important pollinators.

The Cape is also home to another fly with a long proboscis, *Arthroteles cinerea,* in the Rhagionidae family of snipe flies. This species has been observed clinging on to plants even in strong winds,

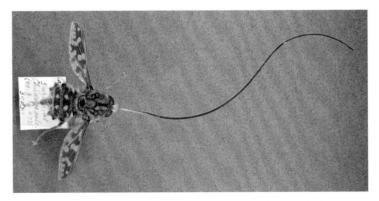

Moegistrorhynchus longirostris, the longest proboscis in relation to body size of any animal – 8 times its body length.

yet it is quite useless at walking due to its poorly formed legs. All of these long-tongued flies (and there are more examples from different families) are crucial for the plants that grow there and nowhere else in the world can you find so many of them.

A species of fly that is less regionally specific is the fungus gnat *Gnoriste megarrhina*, which pollinates the pick-a-back plant, *Tolmiea menziesii*. This plant is a saxifrage, a common plant that was originally native to the North American woodlands. Its flowers dangle down loosely and have long tube-forming petals. The gnat has to reach deep into this flower's corollas, or petals, to obtain nectar and as it does so it rubs pollen onto its body, ready to be transferred onto the next plant. This gnat is but 7 mm (¼ in) long but has a proboscis that is almost the same length again.

Flies are able to cope with such long proboscises because of suction pumps in their heads. Fluid enters the proboscis by capillary action, a process whereby the pressure of cohesion and adhesion causes the nectar (or blood) to flow up the tubes. This can however be very time-consuming, and flies have sped up this process by using suction pumps. Six different types of suction pumps within the head region have been identified, with the type and number varying in different groups. For example, the tangle-veined fly, *Prosoeca* sp., which has a very long proboscis, has two such pumps to assist with nectar uptake, whilst the horse fly, *Philoliche* sp. has just one. The hover flies in the *Rhingia* genus not only have several suction pumps, including a pump at the base of the labrum which generates the pressure to suck nectar into the food canal, but have also cleverly developed a way of protecting their elongated rostrums when not in use, with a very distinctive beak – they look like they have a spout coming out of their heads. Their mouthparts, when extended, are about nine times longer than the beak but are completely tucked away under the snout when they're not feeding. At the tip of their proboscis they have a bristly pair of labella which they use to dab at or scrape the flower to remove the pollen.

Moving away from the warmth of southern Africa, some of the most hostile regions in the world are the Arctic and its polar opposite, the Antarctic, where survival is hard and flies have to cope with extreme fluctuations in temperature and daylight. In the Arctic the average yearly temperature is -40°C (-104°F) and the summer temperatures only reach the balmy heights of 10°C (50°F). On top of that there is very little shelter from the high winds as the landscape is dominated by very low vegetation and there is no tree cover. Only 4,000 or so species of insect have been described from these regions and roughly half of these hardy little critters are flies. And these are of critical importance to the environment through their pollination efforts – bees don't cope well in such extreme habitats and so, thank goodness for the flies. Even though bumblebees have thicker coats they struggle with such low temperatures or high winds and the species numbers are greatly reduced. Flies on the other hand have adapted well to this environment and many of the plants have evolved alongside them.

Adult, non-biting midges from the family Chironomidae are some of the most important pollinators in the Arctic. These flies have had to survive arduous conditions to get to this stage of their lives. They undergo long periods of freezing as well as extremes of light or darkness. *Smittia* is a genus from this family that includes species found all over the world, with several important pollinators adapted to living in very cold conditions. *Smittia velutina* is one of the most dominant species found in the High Arctic and an important pollinator of plants including *Saxifraga oppositifolia*, the purple saxifrage, a dominant Arctic plant species. This plant has been found on Kaffeklubben Island in north Greenland, at 83°40'N, the most northerly plant locality in the world, and also in the Swiss Alps at over 4,500 m (14,760 ft) – this plant likes extremes! Its pollinating midge is an early emerging species and, interestingly, is thought to be parthenogenetic – a form of asexual reproduction where the offspring develop from unfertilized eggs – as no males have ever been found. Parthenogenesis is a particularly useful strategy if you want to produce vast numbers of eggs quickly to capitalize on an abundant, but possibly short-lived, food source. The flowering periods for plants such as purple saxifrage are very short and as such the flies have to develop rapidly to keep up. Why spend time fornicating when there is food to be harvested in a short space of time? An investigation into their ovary development found that the flies were able to mature in about three days, so vast numbers could be produced during favourable conditions. What is rather nice about these flies is that they are diaheliotropic, which means they are sun worshippers and rotate with the sun. Experiments have shown that they rotate around on the flowers to ensure they are exposed to the maximum amount of sunlight as possible. And, a fact that I find most amusing, these flies are what is termed ombrophobic – they don't like the rain! So if it does start raining they use the flowers as umbrellas and hide underneath them.

It's not just midges that are found in the Arctic; there are root maggot flies (Anthomyidae), dance flies (Empididae), frit flies (Chloropidae), dung flies (Scathophagidae), cheese skippers (Piophilidae), mosquitoes and house flies to name but a few. In 2016, researchers from Canada, Denmark, Finland and Sweden published their findings on Arctic pollinators across northeast Greenland. Not only were the flies shown to be so much better at pollinating than other insects but house flies were better than the rest. And within that family one species collected the gold medal as top pollinator, *Spilogona sanctipauli*. These drab little house flies are key pollinators so the recently observed decline in their numbers can only be cause for concern.

Male mosquitoes (and some females) are nectar feeders and we have known that they are important pollinators for more than 100 years. But we've only recently really been studying individual mosquito species, and there are still many assumptions and vague speculations as to their role and importance. We do know that the species *Aedes communis* is an important pollinator of the blunt-leaved bog orchid, *Platanthera obtusata*. It doesn't carry the pollen on its legs or body like other pollinating flies. Instead the pollen balls become stuck to its eyeballs while the mosquito is head down in the plant trying to reach the nectar. The lengths they will go to.

Plants in inhospitable regions have developed little tricks to aid their beleaguered pollinators. As well as providing food, the plants offer warmth and protection. Some plants can raise their own temperature to 15-25°C (27-45°F) higher than the surrounding environment and so keep the snuggling flies cosy, enabling them to warm up enough to fly and so keep pollinating the community. This plant strategy is called thermogenesis and is found in at least 10 families of angiosperms. The skunk cabbage, *Symplocarpus foetidus*, a plant found across continental North America, can raise its temperature to 35°C (63°F) above the ambient environmental levels when conditions become too harsh. This enables the plant to flower while there is still snow on the ground (it melts the hard ground and then the snow) and so benefit from the early emerging pollinators. And the flies themselves have some ingenious coping strategies for the long, cold dark winters and short, only slightly less cold, summers. There may be summer as well as winter cocoons to protect the larvae. Some become tolerant to freezing, like the fly *Heleomyza borealis,* which has larvae that survive in temperatures down to -60°C (-76°F), and others, like the Arctic gall midges, super-cool themselves and so are able to remain unfrozen even in extreme temperatures of below -62°C (-79°F) .

The Antarctic is even more species-depauperate than its northern counterpart. There are very few insect species to be found there and most of the plants have subsequently evolved to be wind-pollinated. Back in 2010, Professor Peter Convey and his research colleagues published an article about two colonizing fly pollinators on South

Georgia, a sub-antarctic island. The arrival of these pollinators was not a positive event though, as it is thought they will encourage other invasive insect species that previously couldn't establish permanent populations there. The larvae of these species are able to consume and break down dead and decomposing material and as such release more nutrients into the soils, so affecting the balance between existing plant species and their environment. It is feared their influence may gradually alter the environment and many of the original, very environmentally specific species will disappear.

There are plants in other regions that are not considered to be insect-pollinated but rely on the wind to transfer pollen. For example, we typically think of grasses as being wind-pollinated as they lack the obvious bright flowers to attract insects. But in fact many types of grass are pollinated by insects. In places like forests, where there is little or no breeze, some species of grass have turned to flies to help them spread their seed. Some of these pollinators are found in the scuttle fly family Phoridae. This family has perhaps the most ecologically diverse collection of flies in comparison to all the other families of flies, and as such maybe of all the insects. Identification of these tiny flies, and I do mean tiny – a recently described species measured just 0.4 mm long – is very difficult though and several genera within the family are numerous, the genus *Megaselia* has over 1,500 species. I call them horrid phorids for giving dipterists taxonomic-induced headaches whilst trying to identify them – most dipterists physically crumple at the thought. Luckily there are a few taxonomic experts out there who do work on this family as they are very important in many ecosystems, including forests. I can attest to the numbers that are found in these environments as a while ago I had a series of pitfall traps (plastic cups dug into the ground) laid out across a rainforest in Costa Rica. These traps when collected were dominated by beetles and these rather strange-looking creatures. They looked like insects in that they had the usual head, body and legs, but they were wingless. It turned out I was being hoodwinked by wingless females of this phorid family. Phorids are dominant in many ecosystems including forests and although information is limited, several species have been found to pollinate the rainforest grass species *Pariana*.

We are now discovering that more and more species of flies are either accidentally or purposefully acting as pollinators, including some unexpected ones. Many species of house fly possess pollen-retaining bristles and we are just beginning to study the impact these species have. House flies and flesh flies are also often tricked into pollinating plants. Many species of fly are not nectar feeders and like nothing better than a piece of meat, the older the better – a nice bit of rotting flesh is the perfect repast. And plants have cottoned on to this and are mimicking the smell and/or the appearance of decomposing flesh. Pollination by flies on rotten meat-smelling plants is called sapromyophily, which means flesh fly loving, and there are many plants that depend on this. In the *Aristolochia* genus of plants, more commonly called the Dutchman's pipe, there are a large number of species that are rather pungent. To the flies their aroma of fresh carrion or dung is nothing less than exquisite. Once lured in, the flowers have long tubes internally covered in hairs which act at times to trap the fly and enable pollen transfer.

This process of kidnapping flies for pollination is at its most sophisticated with *Aristolochia grandiflora*, the pelican flower. The flower has a huge landing pad that acts to direct the flies down into its large reproductive chamber, within which the flies are held captive. On the way down to the chamber the flies brush past stiff hairs called trichomes that point into the chamber and act like barbs, preventing them from getting back out again. But fear not, it is not all over for the fly. This is just the first part of a three-stage reproduction process. Firstly the plant removes the pollen of other plants of the same species from the fly and fertilizes itself. Secondly, over a day or two, its own pollen-producing organs mature, and the fly rubs against this new source of pollen. Then finally whilst incarcerated, the fly munches on nectar produced in the walls of the flower. This stimulates the plants to destroy the thick (gaol) hairs and so enable the fly to escape. Free, but not unburdened, the fly heads off once more to find another sex-starved plant. Genius.

There are many other examples of plants that use so-called carrion flowers (or corpse flowers) to attract flies. One of these is the enormous *Amorphophallum titanum* (titan arum), the plant with the largest unbranched inflorescence in the world, reaching heights of three metres

Aristolochia grandiflora, the pelican flower, with its huge landing pads which direct the flies down and into its reproductive chamber, where it is incarcerated until it finishes its job.

Titan arum, the largest of the 'corpse' flowers, attracts the underworld flesh fly pollinators by smelling like rotting flesh.

(9¾ ft). Imagine if you can an oddly coloured, enormous, half-peeled banana. Titan arum is an unpredictable flowerer, but when it does, it emits this incredibly powerful carrion odour! Some claim it has the foulest odour on the planet. And the flies, flesh flies to be exact, love it (though dung beetles also pollinate the plant). Not only do these flesh flies assist in pollination, their larvae are also parasites of snails – quite a gardener's friend. These plants, and other similarly pollinated plants, have taken this method of attracting the pollinators a step further by heating up the flowers so they appeal even more to the insects – warm rotten flesh. I was recently able to collect flies from a flowering titan at the Royal Botanical Gardens, Kew, in London, and I can testify, quite emphatically, that they do indeed stink.

Flower deception like this has evolved in more than 7,500 species of flowering plants, and two-thirds of these sneaky species are orchids. *Epipactus veratrifolia*, the eastern or scarce marsh helleborine, is a type of orchid that mimics the alarm pheromones of aphids, and the flowers have aphid colouring so as to attract the aphidophagous larvae. The migrant hover fly, *Eipeodes corollae*, is one such species of hover fly that gets fooled by this rather sophisticated trick.

There is still so much to discover. A paper on pollinators published in 2015 by Katherine Orford and colleagues highlights that our knowledge of the pollinators is generally restricted to one or two very well-known families but there are masses more that we don't have any information on at all. It's also important to recognize that pollination isn't the realm of bees alone. In fact, Alison Parker and colleagues at the University of Toronto, Canada, developed a computer model comparing the effectiveness of bee and fly pollinators. They determined that, because bees hoard the pollen they gather, flies, which don't do this, increased pollination events with more visits. It really is time we really started to re-evaluate the role of pollinating flies, if not just for pepper and chocolate.

CHAPTER 3

The detritivores

My Mama always said dyin' was a part of life.
I sure wish it wasn't.

Forrest Gump

THERE IS NO ESCAPING THAT FACT THAT dying is an essential part of life, to clear the way for the next generation. And someone, or rather many ones, have to help with this recycling process. So let's talk about the flies that are detritivores, the gardeners of the countryside, the flies whose larvae munch away the dead leaves and twigs, or feed on the subsequent decaying organic mould. These flies, along with bacteria, fungi and some other arthropods, perform the incredibly valuable role of recycling nutrients back into the environment. And it's not just in woodland or garden communities, but in all terrestrial habitats, including rivers and even beaches. In fact detritivores can be found wherever there is rotting vegetation. I have seen experts in their field, my fellow dipterists, lying fully clothed on a beach, covered in sand, armed not with an edifying holiday read but with a net, energetically sweeping above some rotten seaweed for shore flies, such is their drive to understand the secret life of flies. But don't feel

Adult males of timber flies, *Pantophthalmus* sp. can reach an impressive 8.5 cm (3¼ in) with its wings spread out. Although intimidating, they probably do not feed as adults.

too sorry for us fly-catchers, occasionally that beach is somewhere gorgeously tropical, and I've been unlucky enough to undertake fieldwork under such horrendous conditions as walking along a sun-drenched Caribbean shore sweeping flies off the seaweed. There are many reasons to study flies.

In the previous chapter I discussed how useful flies are as pollinators. But they are perhaps more commonly known for being decomposers – the composters of the animal world, recycling both plant and animal material. It has been said, by me at least, that if it weren't for them we would be knee-high in faeces. So thank you flies for the important but revolting task you peform. All flies that break down waste are called decomposers, but those feeding on plant material are called detritivores, or less commonly saprovores. Determining the specific diet of a detritivore is difficult however, and many may have a more varied diet than we currently suspect. We are still unable, in most cases, to determine whether the flies gain their nutrients from the plant material itself, the associated micro-organisms that are also decomposing the plants, or (more than likely) both of these. For the sake of this book I have classed these examples as one.

In both terrestrial and aquatic ecosystems, most of a plant's organic material remains intact for some time after it has died, which means it becomes available to larvae to decompose when it falls to the forest floor. The larvae break down this material by chewing it into smaller parts – the plant's rigid walls are broken apart and the rich organic contents are made available to the larvae as well as to the wider community. This process of breaking down plant matter into ever-smaller parts enables microbes to get to work on recycling the now available organic cell contents, such as nitrates, back into the ecosystem.

Many adult insects have very different requirements to that of their offspring and as such minimize competition with them for resources such as food or shelter. This has led to different morphological adaptations – for example, to aid optimal dispersal adult flies have developed wings which some have subsequently lost. The adults and larvae of flies, as with other many other insects, have undergone modifications to their diet both across species and across generations.

Looking at insects as a whole group, the mouthparts of the more ancient, or as we like to say more primitive, adult insects, are divided into five component parts – labrum (upper lip), hypopharynx (analogous to the tongue), labium (lower lip), mandibles (jaws) and maxillae (little jaws) – a good example being grasshoppers. Grasshoppers go through incomplete metamorphosis, more often called hemimetabolous metamorphosis, where there is no major change as they develop. An immature grasshopper looks just like an adult as it develops except for the lack of wings, which it properly gets at the final moult. We know that early primitive insects all had chewing mouthparts like those retained by the grasshopper, but many, including flies, have since modified them. In the previous chapter, we saw that, as adults, flies have developed suctorial feeding, but many species of larvae have retained the primitive mandibulate (jawed) shredding mouthparts. This has led to many species having the most vicious of carnivorous larvae that metamorphose into the most passive of plant-feeding adults.

The suctorial mouthparts of adult flies, like those of spiders, cannot deal with solid food and either have to dissolve it first or only feed on liquids, i.e. nectar and blood. When people talk about being bitten by mosquitoes, I find myself getting tetchy. OK don't get me wrong, adult flies can slice you, pierce you, even stab you with some very hardened mouthparts, but they don't have chewing mouthparts so they can't munch on you. I understand that some may feel this is little consolation for being gored and shredded by a horse fly.

Adult flies have, at a minimum, a labrum, a hypopharynx and a labium. Most adults no longer have hardened mandibles and maxillae, the exception being the bloodsucking nematocerans, including the mosquitoes and midges, and some of the lower Brachycera, including

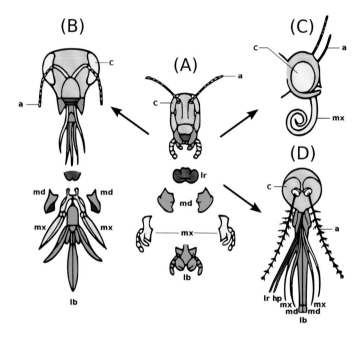

The mouthparts of various insects from the primitive insects with chewing mouthparts like the grasshopper (A) to the suctorial probosces of the bee (B), moth (C) and mosquito (D).

the horse flies. The adults of most species of house flies have the characteristic vomiting mouthparts. This means they have large fleshy pads at the end of their mouthparts that have drainage channels or canals at the base on the underside through which they suck up the liquid. The vomiting is actually the release of digestive enzymes, including amalyase and maltase, which are necessary to break down their dinner into a manageable liquid. This wonderful digestive soup differs to that of the bloodsuckers – but those I will cover later.

Larval mouthparts also vary considerably amongst the flies. As already mentioned, many have retained the primitive, chewing mouthparts – or at least mouthparts that can deal with particulate matter, enabling the larvae to feed on different, more solid food sources. This is an important ecological point because, as mentioned above, it reduces the competition for food with adults over what can sometimes be a very limited resource.

Fly larvae are the dominant detritivore species in most habitats, happy little composters that they are. Whether in streams, on forest floors or in the garden compost, the activities of the larvae (and also to a lesser extent, the adults) are prevalent. Ian Maclean, in his review of beneficial diptera in the *Manual of Palearctic Diptera*, points out that within both terrestrial and aquatic ecosystems, flies are the most dominant in terms of both species numbers and numbers of individuals within each species. But as much as Maclean and other dipterists have written, discussed, debated and (if there is wine involved) shouted about them, flies are completely ignored in most biodiversity studies. This is rage-inducing to the ecologist in me because how can you answer any questions about the habitat if you ignore one of the largest components of it? Even as far back as 1953, Clive Edwards and G W Heath, two British researchers, studied the densities of organisms in the soil of an English woodland. One square metre (yard) had more than 20,000 mites, 15,000 springtails and about 1,300 maggots. That's a lot of space taken up by maggots when you think about the size of them in comparison to smaller springtails and mites.

Decomposing flies are found inside our homes as well. In my kitchen, as in many kitchens, there is a small compost bin. This sits close to my kitchen sink. Individually these are both wonderful habitats for flies and sometimes they become one massive habitat as the bin topples into the sink! These habitats are very appealing to a lovely little detritivore called the owl midge or moth fly of the family Psychodidae, specifically within the sub-family Psychodinae. They have another common name, the drain fly, and most people encounter them near these habitats. To my eye they fly or more frequently walk about in an ever so slightly panicked, jerky fashion that reminds me of someone paranoid about having forgotton to do something. Have I left the cooker on? Did I lock the front door? They are also some of the fluffiest of flies you will ever come across. The adults are small (1–4 mm or ⅟₂₅ –⁴⁄₂₅ in), and their fluff, technically their hairs or scale-like setae, are highly water-repellent. These little creatures are mostly confused with moths but, I think, are much more attractive. Their wings are leaf-shaped and as adults they scuttle about on the surface of the compost, holding out their capes like superheroes.

The family of moth flies (there are six subfamilies) is diverse – in fact many argue these will turn out to be the most taxonomically diverse group of flies as well as the most biologically varied. There are currently around 3,000 described species with some of the more notorious ones being the bloodfeeders in another subfamily called Phlebotominae. Many adults don't feed, or limit feeding to sap or nectar, but across all of the sub-families the larvae are nutrient recyclers. At least two thirds of the described species belong in the Psychodinae subfamily and these are the ones enjoying the slime in your drainpipes. The larvae are completely adapted for this habitat. Depending on the genera, the larvae can be covered with either spines or feathered extensions along the body; so many resemble fluffy hovercrafts floating around on the decomposing waste.

For most of the time the Pyschodines are harmless to humans – in fact many are quite beneficial because they turn the viscous sludge into small, faecal pellets that are more easily dispersed in the environment. But sometimes things go wrong, like too many long hairs blocking the plug hole, and the densities of the larvae build up. Adult physcodines only live for around 20 days – a short time frame but they only need to breed (which only happens once) and then safely deposit their progeny in a suitable habitat. They then take themselves off to die discreetly but are not the strongest fliers and so don't disperse far. The numbers of dying and dead adults

The hairs of the fluffy drain fly of the Psychodinae family are highly
water-repellant, and this allows them to survive in drains, sewers and sinks.

can become so overwhelming – and we are talking thousands upon thousands of individuals – that their bodies pose a danger to our health. Upon death, those beautiful fluffy wings covered in scales and their sensory antennae segments disintegrate and huge plumes of particulates waft around in the air. Workers in sewage farms in South Africa are known to have been affected by bronchial asthma as well as other allergic reactions as a result.

The most commonly infesting species of moth fly found across the world in houses and dwellings is probably the filter fly, *Clogmia albipunctata,* and while this species is rarely problematic, that is not always the case. There have been cases reported of human urinary myiasis in Germany and Egypt, where, upon being admitted to hospital, it was discovered that the patients were passing filter fly larvae in their urine. There is no information about where the patients got their little parasites from, but larvae are commonly seen around the edges of baths and toilets and so they may have picked them up while using these. It appears that the larvae can adapt from a detritivorous diet to a more varied one, such as the one provided by humans. Larvae have been recorded discharged alive in the urine of one of the female patients – another example of how extraordinarily hardy this stage of a fly's life cycle can be.

There are many sources of lovely decomposing plant substances found in more natural settings such as trees. Flies act as decomposers in sap runs (where the bark has been damaged and the sap flows externally), rot holes (where the tree or part of the tree has rotted and a hole appears), under bark and in the heartwood. In the UK some 400 species of fly develop in sap runs and dead wood, which is as much as 6% of the described UK fly fauna. These dark, dank environments are often rich in dead and decaying organic plant matter, as well as some other animals that had the misfortune to fall into them. This rich source of dissolved food is a great resource for many dipteran larvae, and a complex community of decomposers, as well as large carnivorous larvae that prey on them, can develop. These are some of the smallest and rarest and often very temporal habitats. The flies that live in them also tend to be some of the rarest recorded.

When deciduous trees are injured or succumb to the inevitability of death they become homes to the splendid *Ctenophora* genus of crane flies. There are nearly 40 species of these found globally and they are all magnificent in form. They are wasp mimics and are conspicuously coloured with various combinations of red, yellow, orange and black. It's not just the colouring that is so spectacular about the adults, the males also have extraordinary antennae. They are pectinate or comb-shaped, with long spines on one side – it looks like the male is flying around with combs on his head – and they have led to the species sometimes being referred to as comb-horned crane flies. When species live in temporary or dispersed habitats, the male needs as much help as he can muster to find a member of the opposite sex. The antennae are one of the main sense organs for flies and they use them for touch, taste, hearing and, most importantly for these dispersed species, smell. The antennae are covered with sensitive hairs that are able to pick up mechanical and chemical signals. To increase their effectiveness males often have very plumose or multiple-branched antennae to increase the surface area and so support a greater numbers of hairs. *Ctenophora* are generally confined to orchards and old woodlands where there is a continuous source of dead wood for their larvae to develop in. Their offspring have very sturdy jaws, which they use to munch away at the heartwood. Sadly dead trees are often cleared away because of modern woodland management practices and many small orchards are disappearing. These species, because of their conspicuous markings, are very easy-to-spot indicators of this type of habitat.

Another group of flies from the *Rachicerus* genus in the awl fly family Xylophagidae have equally ornate antennae. *Rachicerus* is the only genus across the whole of Brachycera to have such pectinate antenna. With maybe the exception of the soldier fly, the genus

The extravagantly ornate antennae of the wasp-mimic male crane fly, *Ctenophora flaveolata*. The plumose antennae increase the surface area of sensitive hairs which pick up chemical signals from females.

Ptilcera also has slightly pectinate antennae but nowhere near as resplendent as *Rachicerus*. It does seem odd that these are the only ones in the whole of the brachyceras that have such ornate ones. As with the comb-horned crane flies, they are associated with dead and dying wood. But then so are many other species from this family. Why have the four species from this genus developed such ornate antennae when 131 species haven't? It is a mystery. Personally, if I were a female fly, I would find them most attractive and maybe the female preference is why this group has them.

Crane flies and awl flies are not the only charismatic flies to be found around decomposing wood. Hover flies are conspicuous in this habitat, with many of the males being very territorial over their chosen log. If you slowly approach a hovering male he may come and check you out. And I've found that if I lift my hand up they will often take the opportunity to sit on it.

The golden hover fly, *Callicera spinola,* is aptly named and is one of the largest and most colourful of the British hover flies though it is not very common. The larvae are saprophagous – they feed on decaying wood, and are confined to rot holes in beech trees. While the larvae are dependent on the decaying parts of the trees, the adults have been seen to feed on ivy flowers found on or near the trees. The adults and offspring for once are not so widely dispersed. This species is only found in parts of eastern England, but go north to the highlands of Scotland and you will find another hover fly, the aspen hover fly, *Hammerschmidtia ferruginea.* Its larvae live and feed on the decaying oily layers of – you guessed it – aspen trees. Although found throughout the UK, aspen trees are most common in the north and west of Scotland. Once more, over-zealous management removing the dead trees and over-grazing of the younger trees by mammals has resulted in there being fewer and fewer piles of dead aspen around. You may be unsurprised to hear then that this fly is also very rare. But ongoing research into all of these flies and their environments has enabled us to understand their habitats more clearly, and their numbers are now on the rise. This is a UK example but thankfully it is happening all across the globe.

We couldn't discuss rare decomposing flies without a mention of the Axymyiid flies from the Axymyiidae family. This is a family with only eight species, which may seem small but not the smallest. In fact there are quite a few families that only have one species including the families Oreoloptidae, Evocoidae, Inbiomyiidae, Apystomyiidae and Mormotomyiidae. The Axymyiidae have no obvious common name as such and no-one has really done any real research into them – in fact, we don't even know how they fit, in relation to other fly families. One minute we think they are closely related to one family and then the next we change our minds as different researchers place different emphasis on different body parts, and they are once more moved around within the taxonomic tree. Some authors have placed them in their own infraorder Axymylomorpha (usually a group of similar families but in this case consisting of only one family), while others have them in the Bibionomorpha, which contains the March flies and fungus gnats. They mostly resemble the adult March flies – chunky and robust – but their larvae are very different. Adult axymiids have very reduced mouthparts so we can assume that they don't feed, but as they have been very rarely caught let alone seen, we have to just make an educated guess.

The larvae live in running water habitats, such as streams and springs, inside partially submerged wooden detritus such as fallen branches. The current assumption is that they are feeding on this decomposing substrate and its associated fauna. Available air in the waterlogged environment is not abundant and so they have developed a siphon – a particularly long 'tail' that ends in a pair of spiracles. This retractable siphon has a hardened tip to enable it to be pushed through wood to the outside environment. These larvae also have what looks like a white kelp forest growing from their posterior region! Recent studies have started to look more closely at these flies and we are now learning that, within their specific habitat, they are not as uncommon as we once thought – their larvae can be found in large numbers so dipterists are trying to track down their elusive parents to study their life histories. However this fly was described back in 1921 in the Appalachian Mountains, USA, so in nearly 100 years we still haven't properly found out anything about them.

There is no mistaking some of the truly spectacular beasts of the dead wood fly community – the giant wood flies, timber flies or more simply the 'pantos' of the Pantophthalmidae family. In terms of species numbers, this is another small family, only 20 species, but when it comes to their individual size they are far from small. In fact they are some of the largest flies on the planet – *Pantophthalmus bellardii*, with its wings spread, can be up to 8.5 cm (3¼ in) wide. These flies scare the bejesus out of people when they see them (even when dead and pinned in drawers) as they look like horse flies on steroids. But unlike their bloodsucking cousins, it is thought that these only feed on wood. Considering their size and that we know where they tend to make their homes, the fact that we do not really know much about their life histories, especially what the adults and larvae are feeding on, is ridiculous. The adults don't look like they have functional mouthparts – a strange beak-like object on their face, yes, but we don't think that this is for feeding. One suggestion is that its actual purpose is to open up rotten trees to lay eggs in. Now it's not uncommon for the adults not to feed, and this may also explain why we rarely see the adults as non-feeding adults can be fairly short-lived. Many people believe the larval diet must consist of fermenting sap, while others think it is a mixture of wood (either in the process of dying or dead) and micro-organisms. One of the problems we have in determining the diet is that the larvae are wood-borers and inhabit galleries that are carved horizontally into the tree (dead or living depending on the species) so we can't easily peer in and see what they are doing! Recently researchers have started using new molecular techniques to analyze the gut contents of insects, which will help us ascertain their feeding preferences. Hopefully we will soon have an answer to these long unresolved questions.

Someone who has done research on them is Manuel Zumbado, an entomologist in Costa Rica, who states that pantos seem to prefer mucilaginous trees, the sticky and slimy trees such as kapok, or heavy, sap-producing species such as figs. He describes how noisy these little critters can be – several hundred may inhabit a trunk at any one time and their munching can be heard from several metres away. The larvae have very robust head capsules and massive mandibles to enable them to manoeuvre through (and we believe consume) such tough substrates. Overall they are some of the largest larvae I have seen of all insects, let alone flies.

Another great group of decomposing flies, with some of the most bizarre modifications on their heads, are from the genus *Phytalmia* in the fruit fly family, Tephritidae. It is the adults not the larvae that have the modified heads which have resulted in them being referred to as antler flies. Never has a name been more apt. The males of these crazy flies have amazing protrusions that superficially resemble the antlers of deer. And they use these antlers in a very similar fashion –

The timber fly, *Pantophthalmus* sp. looks like a horse fly on steroids. But unlike its bloodsucking cousins, it only feeds on wood – we think.

to spar with their opponents and show off to the females! The males also use their antlers to defend territories that have the best rotting sapwood which, surprise, surprise, are also the larval habitats sought out by the females. In the signal fly family Platystomatidae there are many species that have widened heads or eyeballs on stalks which, through head-butting and fisticuff fighting, they use to assert their territorial dominance.

Some species of decomposers are not so fussy about their food or habitat requirements and are thus not so restricted in where they live. This is especially so of the aquatic decomposers – the filterers of the waterways. One of the best at doing this is the non-biting Chironomidae midge family. There are about 20,000 species globally and they turn up practically everywhere. We have already discussed the marine species, but there are ones that have possibly established

The antler flies from the genus *Phytalmia* have amazing deer-like antlers which they use to spar with and show off to females.

themselves in even weirder places. Two species have been recorded in Antarctica and are, as such, the largest purely terrestrial animals on this continent. That they can lay claim to this title is even stranger when you consider that these 'enormous beasts' are actually only 2–6 mm (²⁄₂₅–¹⁄₅ in) in length. The larvae of one of these species, *Belgica antarctica,* are omnivorous, feeding on algae and moss but also decaying vegetable matter, and micro-organisms and they construct special winter cocoons to keep warm. These hardy larvae also contain the sugars trehalose, glucose and erythritol, which help prohibit ice from forming within their cells. Most of their life cycle is spent as a larva which can typically reach a two-year life span, while the wingless adults are short-lived, often surviving less than 10 days.

It's not just freezing temperatures that flies can endure. They will go to incredible depths. Lake Baikal, in southern Siberia, Russia, is the largest and deepest freshwater lake in the world at 1,741 m (5,712 ft). The midge genus *Sergentia* has many endemic species in this lake, one of which is found at the greatest depth of any aquatic insect. Midge specialist Anna Linevich spent years studying the lake and back in the early 1970s described specimens of *Sergentia koschowi* found in the fine 'oozes' at 1,360 m (4,462 ft) below the surface. These larvae develop in temperatures between 3.4 and 3.6°C (38.1 and 38.5°F), in borderline anoxic (oxygen-starved) environments. Their conspicuous red colour is due to the presence of haemoglobin, which is easily visible through their translucent bodies. Haemoglobin is a respiratory pigment that is able to store oxygen, and this haemoglobin has a much greater affinity for oxygen than human haemoglobin, enabling the midges to live in what appears to us a very inhospitable habitat. Flies are resilient if nothing else.

The larvae of different midge species vary in their diet but mostly they are decomposers, feeding on particles of organic material. Some prefer small particles that just waft by in the water, while others create their own by shredding bits of dead wood and leaves. By knowing which chironomids live in a given habitat we can have a good guess at the water quality as many species, being filterers, are brilliant indicators of pollution. A very conspicuous filter feeder, *Chironomus plumosus,* is often found in polluted water bodies. Its larvae have bright red bodies, making them a favourite of fishing folk to use as bait. These species are called bloodworms and can be found in densities of 100,000 per m^2 (over one million per square foot). In lakes they are often the most dominant species in the banks and beds, and their mass emergence as adults has resulted in some visually arresting displays similar to that of murmurations of starlings. These can be problematic if the densities are enormous and, as with the owl midges, they can cause respiratory problems in humans. Although at times irritating for us, they are essential to lake food webs, with many fish and birds being dependent on them.

Haemoglobin isn't the only adaptation that helps midges live in poor quality habitats. Some also have a clever way of collecting their food. Many species live in tubes or burrow down into the mud below or on the margins of river beds, only extending their heads out to feed. Some however, including the bloodworm, cast out a sticky net from their burrows that traps passing particles of food. They can construct these mostly conical nets from their salivary silk glands in less than 40 seconds. Like a fisherman, *C. plumosus* casts its net to collect its food, but when it draws it in again it eats the catch, net and all – not such a common practice among fishermen.

Returning to terrestrial habitats, there is another family of flies that is worthy of a mention amongst the great devourers of decomposing material. Drosophilidae or vinegar flies (often incorrectly called fruit flies which are in fact a different family) are some of the most important organisms on the planet. Thanks to their propensity to breed at a rapid rate and their willingness to live in laboratories, we have used and abused them for over 100 years to help us understand genetics. As an undergraduate I was fascinated by images of legs grown out of faces, eyeballs on legs, and genitalia where they really shouldn't have been as examples of how we manipulate gene expression. Back in the wild, the Drosophilidae larvae eat a wide variety of material, but most feed on organic substrates derived from rotting or fermenting vegetation, plus its associated fungi and micro-organisms. Rotting and fermenting fruits and yeasts you say? Yes, these flies are the alcoholics of the insect world! Both the larvae and the adults are known for their love of an alcoholic drink and can consume vast quantities of decomposing material along the way.

We have used their love of a wee tipple to learn more about human behaviour and gene expression. When the adult flies commence drinking they become clumsy and start lolling around, falling onto their backs. What's more, as they drink more they become more amorous and far less able to pick a suitable mate. In the wild these species are heterosexual, but during a drunken laboratory session absolutely anything goes. But as their flirting increases, their ability to perform decreases correspondingly – as Shakespeare writes in Macbeth: Act 2, Scene 3: 'Drink sir, is a great provoker of three things… nose painting, sleep and urine. Lechery, sir, it provokes and unprovokes; it provokes the desire but takes away the performance!' This behaviour is amusing but what does it really tell us? Well, we know humans act in a similar fashion under the influence and using the flies as models we can start to understand the mechanics of this behaviour. With the flies we are able to switch their genes on and off and see what is specifically affected by alcohol and we can use this to understand alcoholism and other related conditions.

There are estimated to be well over 4,000 species of these vinegar flies globally, but the best place to study them in the wild is Hawaii. There are probably more than 700 endemic species on these islands alone, although only just over 500 have been described so far. Recent studies have shown that a single ancestor species turned up around

25 million years ago on this, still very geologically active group of islands. The original species, belonging to the *Drosophila* genus, split into two with the evolution of a second genus, *Scaptomyza*. The former are all saprophagous species, associated with at least 40% of the islands' endemic plants, while the latter are more cosmopolitan, and include saprophagous, parasitic and microbe-loving species.

Common among all detritivorous species is their dependency on gut microbes to help them break down decomposing plant tissues. Interestingly these microbes may have also exerted pressure on the flies to speciate, enabling them to feed on different plants and so reduce competition as closely related species have a very different gut flora. Had Darwin been able to look at the microbe and fly communities of these islands I am sure he would have found them far more interesting than the finches of the Galapagos.

On the subject of beautiful island ranges, go to any beach and look at the strand line covered with seaweed. Take a moment – maybe lie down parallel with it and keep your head near the ground – look over the seaweed and you will see lots of insect activity. Ray Ingles' *Guide to the Seashore*, in common with many publications of its type, states that very few insects have adapted to coastal environments. My copy only mentions two species of fly – *Coelopa frigida* and *Coelopa pilipes* from the seaweed or kelp fly family Coelopidae, which is better than most seashore guides that often ignore flies completely. But this is not an accurate reflection of their presence as many flies have managed to adapt to decaying seaweed and flotsam zones on the beach, including the kelp flies, and if you look carefully amongst the seaweed you can see them munching away!

Kelp flies are a slightly odd-looking family of flies. Their shape is like that of a normal fly that someone has stood on, flattening it. And their legs are completely covered with spines. To the untrained eye many of the 40 species or so look alike, with their long wings held precisely over the back of the abdomen, and their slightly flattened heads and bulbous eyes. By flattening their bodies they are better adapted to squeezing through and between strands of seaweed. These flies feed on slimy masses of decomposing seaweed which is such a rich food source that they can reach very high densities given the right conditions. Their fondness for breeding allows us to culture them in laboratories and use them – as with *Drosophila* – to study genetics and ecology.

For many species of animals the quantity of food available affects how big the males get (females too, but they're generally not the fighters so don't need to be so big and impressive). Most of the time the change in larval food quantity results in small changes in body size, but shore flies can vary greatly. Derek Dunn, a biologist specializing in mating systems, and colleagues describe the mating habits of these flies as a 'scramble competition' in which the males that are the largest, and so best fed, fight most aggressively and are able to mount more of the females. But the larger females are also able to fight back to remove unwanted suitors – hence explaining why they are variable in size.

These flies do not only feed like they are at a medieval banquet during the larval stage. The adults also know how to sniff out good quality food sources. This has led to some strange inland invasions. There are many intriguing accounts of these flies invading seaside chemists, watch repairers, perfumers and dry-cleaners. This is because the adults are attracted to trichloroethylene, a chemical used for cleaning delicate products as well as a base in perfumes and solvents. Although not a naturally occurring substance, it has quite a sweet odour and smells similar to decomposing seaweed – to these flies that is – and the flies swarm to these shops confused by the smell of the chemicals. To me rotten seaweed has a distinctive odour and I, for one, am grateful that so many flies are working on removing it.

The kelp flies are unusual in that they are all decomposer specialists on beaches – working away in these habitats and nowhere else. What happens in the seaweed stays in the seaweed! Kelp flies, unusually, are restricted just to temperate coastlines – other families of seashore flies dominate in the tropical regions. In the moist seaweed there, families present include dung flies (Scathophagidae), root maggot flies (Anthomyiidae) and black scavenger flies (Sepsidae). These fly communities are so good at feeding that they are able to break down the kelps into soluble units small enough to be washed back into the seas in the short interval between tides.

Shore flies of the family Ephydridae like the seaside as well but they have also been found near lakes. Some of these detritivorous genera are known as brine flies (*Ephydra),* as they live in very alkaline or salty environments. For example *Ephydra hians,* the alkali fly, is found throughout the northern Americas and its larvae can be observed feeding on detritus and algae on salty lake bottoms. They do not even have to ascend for air as they get all their oxygen from the photosynthesizing algae. Even more unusually, adults have been observed underwater. They are able to descend with 'air tanks' – pockets of air trapped by the hairs on their legs. These flies can also be found in high numbers. Lake Mono in California is famous for these swarms but, thanks to their air tanks, they can also feed below the surface. In his 1872 book *Roughing It* Mark Twain describes holding them underwater with harsh and detrimental consequences and how, upon being released, they 'walk off as unconcernedly as if they had been educated especially with a view to affording instructive entertainment to man in that particular way.'

So whether it's in the bin or on the beach, the actions of our vegetarian decomposers are incredibly valuable. Plant cell walls are really difficult to digest and even after death do not readily disintegrate and so we need our little detritivores to once more release the essential nutrients contained in them back into the environment. Long live the decomposers!

CHAPTER 4

The coprophages

Kings and philosophers shit – and so do ladies.

Michel de Montaigne

T HE DECOMPOSITION of plant material is an obvious but often overlooked process, a tad dull if essential, but it's the decomposition of animal waste that gets everyone talking. Jean Henri Fabre (1823 to 1915), a whimsical French entomologist, who many regard as one of the fathers of modern entomology, wrote often about flies, those neglected beasts, in this regard saying 'you think they are horrid, dirty insects; but they are not; they are busy making the world a cleaner place for you to live in.'

Flies and indeed all creatures that feed on animal waste are termed coprophagous – from the Greek *copros* meaning faeces. Both adult flies and larvae have been recorded on and around dung but generally it's the more advanced larvae, evolutionarily those that we call maggots, which have adapted to feeding on this delicacy. Plop (pun intended) yourself down in front of a good dung pat and you can enjoy watching the adult flies flirting with each other. And why wouldn't they choose to leave their offspring there when you

The holotype - the specimen on which the description and name of a new species is based - of the terrible hairy fly, *Mormotomyia hirsuta*, a most aptly named fly.

consider the nutritional value of such eutrophic substrates? They are little nutrient nuggets.

There are many species of fly associated with the breakdown of this abundant material, some of which we consider to be a nuisance when there are too many of them. But think about it, very few things don't defecate – it is an incredibly reliable, if ephemeral, food source. All animals defecate even if it's sometimes restricted (some larvae such as those of antlions, bees and ants don't – they hold on to internal waste products until they pupate as they are stationary throughout the larval stage and defecating would result in a rather messy home). Pixiu, the Chinese dragon of wealth, is less fortunate, and when he broke the law of heaven by defecating on the floor, he was punished by having his anus sealed – luckily no real animal has suffered this fate. If you search for defecation rates on the internet – I have – you'll find that scientists have been researching these for a while and in subjects ranging from agricultural animals (cows at 26 kg/57 lb a day) to brown bears (3.3 scats a day) to elephants, which egest a phenomenal 100 kg (220 lb) a day, nearly 40 tonnes a year! Humans on average, egest 1 kg (2 lb) a day. You may be glad to know that globally defecation rates in the outdoors (rather than in toilets) have dropped steadily over the past 30 years due to increased sanitation facilities, but faeces still end up somewhere and need to be dealt with.

The story of dung is not always straightforward, as exemplified by the introduction of cattle to the wilds of Australia. Before this the decomposer community had been getting along just fine, with kangaroos and other large creatures native to Australia having co-evolved alongside them for millenia. Some decomposers had even evolved very intimate relationships – clinging to their marsupial hosts until these hosts defecated and then dropping off so as to be first in line for the feast. Now cows, as I'm sure everyone knows, produce large, watery pats at, as we have established, a fairly rapid rate. Before cows were introduced to Australia there were no specific dung beetles or other insects adapted to cow dung as there were no animals similar to cows in Australia. Australia was dominated by non-placental mammals such as marsupials, with the exception of rats and bats. Humans came along and changed all that, introducing 13 species of ungulates including camels, pigs and, most importantly, the aforementioned cows. An incredible figure of 12 million cowpats alone are deposited by 30 million cows every hour! No beetle 'cleaners' for their faeces, which is of a very moist consistency in comparison to the fibrous variety, led to a potentially huge, sticky and smelly mess for the farmer. Thankfully, other opportunistic species moved in to do the job. The bush flies, *Musca vetustissima*, a relative of the more common house fly, *Musca domestica*, ate and copulated with abandonment, gorging on this new, very rich food source. And, from the 1880s to the 1960s they rampaged across the landscape and were not only problematic for the cows but humans too, due to their vast numbers. The Aussie wave or Aussie salute comes from these flies as people swat them away from their faces. And who can forget the cork hat – a shearer's 'must have' – brought about by the massive densities of flies! Luckily, after a few false starts, scientists were able to find dung beetles that could be moved to Australia to recycle the dung and reduce the health hazard.

Although problematic for many years in Australia, *Musca vetustissima* is nothing in a global context when compared to *Musca domestica* which is often cited as one of the animals that threatens human health the most. But it is the habits of humans that have encouraged this relationship. These flies have probably the most commensal relationship – a relationship where one side benefits and the other is unaffected – with humans; they go everywhere with us because we are sloppy creatures leaving waste of every conceivable type around us. There are many species of house fly, over 4,000 have been identified so far, but actually very few have any real interaction

An inverted eye lid showing a face fly, *Musca sorbens*, feeding. Face flies also love feeding on cow and dog faeces.

with us. Most get on with their lives and ecological roles such as decomposing waste, so we should be less antagonistic towards them.

From my perspective as an entomologist, house flies are attractive creatures with large, bulbous eyes, often reddish in colour. They have some of the most bristly faces out of all the flies, with hairs appearing to emerge in a haphazard fashion all over them. Do not be fooled however as these bristles are anything but random and are of great taxonomic importance as their angle or position can define a species. House flies can travel vast distances, some have been tagged with radiation by scientists to determine how far they go and have been discovered 16 km (10 miles) away from the original tagging site. In moving from one food source to another, they can transmit any pathogens they pick up, as well as leave dirty footprints. For a species that does interact with us, that becomes a problem. *Musca sorbens*, the ominously named face fly, loves nothing more than feeding on cow or dog faeces. But as their name suggests they also love feeding on the fluid of human eyes and our nasal discharge. Now that's not a lovely combination and it's obviously one that worries health officials globally. And our tendency to keep livestock near human habitation only increases our contact with these flies and any pathogens they may be carrying.

Blow flies, along with house flies, are often found on humans and our waste, and they too have been implicated as pathogen transmitters. Back in 1929 in the slums of China a study found that two species of blow fly, *Lucilia* sp. and *Calliphora* sp., were found to have on average over 3.5 million bacterium per fly. Sewage in water systems is a major source of bacteria, where algal blooms, as a result of excess nutrients in the sewage, provide a rich food source for the harmful bacteria *Salmonella* and *E. coli*. Too much enrichment can also starve the waters of oxygen and so kill off the existing fauna and flora.

The traditional way of dealing with any manure from agriculture was to hurl it around our fields as slurry, which often resulted in a very smelly bus journey to school for yours truly, but nowadays our production of agricultural manure way exceeds this use. *Musca* species are often the only creatures around able to recycle the waste and deal with such vast quantities of it. Instead of letting these quantities intimidate us we have started thinking about how to use them to our advantage. Apart from animal waste, one of the many issues facing farmers today is having a reliable, good quality food supply that is affordable. Researchers are now exploring the feasibility of letting the maggots consume the animal waste – manure – and then using these maggots to feed back to the livestock.

House flies are not alone in appreciating faeces and helping us to deal with it. There are 30 fly families that include waste decomposing species and 13 of these are reliant on waste as their main food source. The true champion of dung consumption is the impressive black soldier fly, *Hermetia illucens*, a species from the soldier fly family Stratiomyidae, a dashing figure of a fly with a metallic exoskeleton. Soldier flies are so named as they are thought to look as if they are wearing military uniforms. In the UK the smaller soldier flies are referred to as the majors and then, as the species increase in size, they are promoted to colonels, then brigadiers and finally the largest species are the generals. The adults are often striking – some have bands across their eyeballs, whilst others have metallic bodies and yet others have very large derrières – which always results in me singing Queen's 'Fat bottomed flies [in place of girls] you make the

The industrious soldier fly, *Hermetica illucens*, one of the best recyclers around and the possessor of a metallic exoskeleton.

rocking world go round', when I see them in the Museum's collection. Adult black soldier flies do not consume much but instead leave that responsibility to their larvae. And by gum can these larvae consume and organic waste, such as manure, is their delectation.

The soldier fly is incredibly easy to breed which has led to there being a very active community of black soldier fly (BSF) breeders and buyers. Commercial and domestic factories have been set up globally to supply both the sewage treatment industries and the livestock production industry – you can even purchase a mini waste factory for all your domestic waste needs (well, maybe not all). The idea is that alongside a chicken factory/farm, you have a factory producing maggots, and the second factory's inhabitants consume the waste of those in the first factory. Then, twisting it around, on pupation of the flies, the first factory's inhabitants then consume the second ones! A very neat little system but slightly unfair on the poor flies as they do all the work, their only reward being a premature death.

What is it about this specific species that makes them so good for both waste disposal and domestic animal feed? For one, these little creatures grow incredibly quickly. It takes just four days to develop from egg to larva. They are also one of the most efficient nutrient converters – converting waste into protein – of all the insects. The BSF larvae and pupae are eaten whole by the hungry chickens and these morsels pack 42% protein and other essential nutrients such as calcium and amino acids. There are trials in the USA where they are feeding freshwater prawns on BSF and apparently the only difference between them and your regular prawns is in the colour of the prawn – humans wouldn't be able to tell the difference otherwise.

You may not like the idea of consuming insect meat as whole animals, but if it is broken down and converted into products such as flour or protein pieces then we can easily add it to our diet and our livestock feed. (We wouldn't necessarily eat insects that have been produced by animal waste as some people may object!) We can't catch any diseases from eating insects as their own pathogens are adapted to them, not us. It's easier for us to catch something from pigs, cows and chickens, because we are very similar, in terms of fat, moisture and temperature to our edible mammals, and so disease can spread more easily.

Not only do BSF remove waste and subsequently provide a food source, they also prevent the spread of disease. In keeping the level of faeces low in comparison to traditional farming practices, they are limiting the spread of other pathogen-carrying, dung-loving flies. The BSF are indeed exceptionally helpful little creatures. We use them now in fish farms, and so they may also help the world's seas from being over exploited, and as pet food for the more exotic pets. 'Give him the best' is the gleeful tag line for the Phoenix Worm website which, and they state 'recommended by veterinarians and used by major zoos', sells BSF for a variety of animals.

But let's not just think about agricultural and domestic waste. What about all the other animals out there defecating? Bats defecate no surprise there. And in a cave this faeces – guano – can build up into vast heaps such as with the guano mountains on the Chincha Islands off the coast of Peru, where deposits of cormorant guano have reached a height of more than 50 m (164 ft). As with all faeces, it is a valuable resource, a veritable feast of nutrients, and many strange looking diptera are found in these environs.

Mystacinobia zelandica is one of them, a species of fly that lives on bats and is only found in New Zealand. That's not too uncommon, as there are many species of fly only found in New Zealand, but this is a very odd species in terms of its taxonomy, its biology and its behaviour – even the way it was discovered is unusual. It was described back in 1976 by Beverley Holloway, a New Zealand entomologist. When first coming across this species, it must have been difficult even to establish it was a fly as it lacks, or has modified, several key morphological features so it really doesn't look

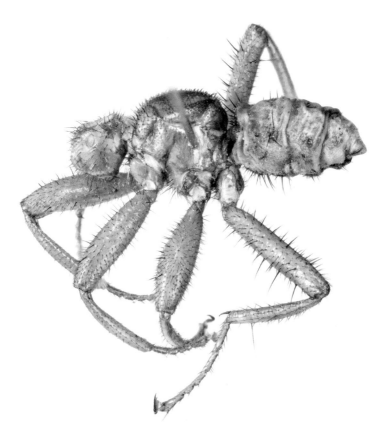

The rather odd-looking *Mystacinobia zelandica*, the New Zealand bat fly, with claws on the end of their feet to grip on to and move through the hair of bats, their hosts.

like a fly any more. Both male and female adults are wingless with long spindly legs, resembling those of a spider, and they have large claws on the end of their tarsus (feet), enabling them to grip on to, and swim through, the hair of their bat hosts. As with many animals that live in the dark, these flies have tiny eyes that are of limited use.

These flies were first documented a decade before Holloway named them. Peter Dwyer, an Australian zoologist working with the New Zealand lesser short-tailed bat, mentioned in his paper the flies that he collected from two of the three species of New Zealand bats. He stated that a Mr H. E. Grubner had recovered some very odd-looking creatures from a captive bat (a bat that subsequently escaped). Initially they were thought to be ectoparasites, a species living and feeding in a way that was detrimental to its host, but no other ectoparasitic species of fly had been found on any of the New Zealand bats. These bats roost up massive trees, which meant that opportunities to study them were limited. But luckily for Holloway, and science, she had a lucky break when a tree housing a roost of lesser short-tailed bats collapsed, and suddenly the researchers were able to gather flies before the bats (and these flies) had moved on. And she was able to catch enough flies to determine two things. Firstly, gut analysis revealed that the flies were not seeking a blood meal from the bats as their stomachs contained guano not blood. They were not ectoparasites after all. In fact the bats benefitted from this relationship, with their surroundings being kept clean by the flies. Secondly she was able to establish some taxonomic resolution. These flies were not from the bloodfeeding bat fly families already known but instead from a completely new family. This one species of fly, *Mystacinobia zelandica*, is the only living representative of its genus and its family Mystacinobiidae.

The adults are described as physogastric – that is, they have massively swollen abdomens which resemble those of queen ants and termites, and also have massively enlarged ovaries and testes to enable greater reproductive rates. Once the eggs have hatched, larval development is fast – it is thought they develop in only 20 days, the larvae growing and feeding in the guano. What is really amazing about these flies, in comparison to not only flies but most other insects, is that they groom not only themselves but each other!

In fact they hang around in little social groups. This is essential as guano is, as I am sure you can imagine, quite sticky and the flies can only really clean their own head and front legs. They need their companions to help clean the rest of them. This is not just restricted to the adults as the larvae have been observed cleaning the adults as well. I know of very few truly communal groups of flies. And just when you think it couldn't get any weirder Holloway states that the males were also heard 'zizzing' – I don't think I had come across that as a vocal description before but you get its meaning. The males vibrate, which in closed environments sounds 'zizzy'. This is thought to act as a deterrent to predators as the males (and to a lesser extent the females) do it to protect their offspring. This is the only case in the diptera that we have discovered to date where there is anything close to a social species in the true sense.

There is another family of flies that we believe are also guano feeders. They live in caves in eastern Kenya and they may be even rarer than the New Zealand bat flies. *Mormotomyia hirsuta* is a species that is the sole representative of the genus *Mormotomyia* and family Mormotomyiidae. People had thought that this fly was extinct and there is very little evidence of them in any museum collection. The Natural History Museum in London has just a couple of specimens that were collected in 1933 and 1948. Until relatively recently species were described solely on their morphological characteristics (we now use genetic information as well), which really didn't help much when it came to these creatures. Not only were they wingless and spindly, they were also incredibly hairy, a feature that earned them the moniker of the terrible hairy fly, a straight translation of the scientific name. The really hairy ones are the males and they are also generally much larger, with much longer legs than the females. This species didn't appear to have many features that could link it to others.

When *Mormotomyia hirsuta* was described back in 1936, Edward Austen, the English dipterist responsible for describing it, tentatively

The head of the terrible hairy fly, *Mormotomyia hirsuta*, thought to be one of the rarest creatures in the world.

placed it near the Heleomyzidae, in the Sphaeroceroidea superfamily, but others placed it in the Hippoboscoidea superfamily, a very different group. No taxonomic resolution could be made and, as the specimens were old, many of the molecular techniques could not be applied to resolve this taxonomic anomaly. Numerous expeditions back to the original caves were made to find fresh specimens of this elusive fly, but all were unsuccessful. So the specimens have just sat there, looking weird in the drawer. Until very recently that is.

In 2010, and on a subsequent trip to Kenya in 2011, three researchers, Robert Copeland, Josephat Bukhebi and Ashley Kirk-

Spriggs rediscovered the species and are now beginning to unravel some of the mysteries surrounding it. And everyone in the fly world is very excited about it (they are…honest!). Both traditional media and the electronic world were a flutter with this little beastie. It is not often that I do this, but to quote the *Daily Mail*: 'It went missing for 62 years, but now Africa's "terrible hairy fly" has been discovered in remote caves in Kenya by a couple of intrepid dipterists'. That's a tautology – all dipterists are intrepid!

As with the New Zealand bat flies, these were originally thought to be ectoparasitic flies but they are now believed to feed on guano. New specimens enabled scientists to undertake both morphological and molecular analyses, and the terrible hairy fly is now believed not to be in either of the previously proposed superfamilies but in the Ephydroidea, a superfamily that lives between the two and includes shore flies and vinegar flies. It is so exciting when there are finds like this enabling us to answer long-established taxonomic puzzles. And knowing that this species is not extinct is even better! Originally only found in a very restricted locality, the researchers are now studying the fly's life history in the hope of determining its habitat requirements and so locating similar sites elsewhere in Kenya where other populations may live. This species, as it currently stands though, is thought to be one of the rarest in the world but it is not yet on the International Union for the Conservation of Nature's Red list of the world's most endangered creatures – not even its fluffy physique has endeared it to the world enough for people to rally to its cause.

Moving away from bats, there are many other families of flies that specialize in feeding off different types of animal faeces. Did you know for instance that there is a wombat fly? There are many wombat flies – so far in fact there are 24 species described from the family Heleomyzidae that all feed on wombat dung. The family is quite a hodgepodge when it comes to feeding preferences but many of them have a taste for waste. And wombat dung is not your regular pat-forming dung. Instead it is deposited in nice little cubes. Now I know what you are probably thinking – how does the wombat pop out square poo? Do they have a sphincter muscle (a band of muscle) that is pulling it into shape? We don't know the answer to that question but we do know that it has an extraordinarily long digestive tract. It can take weeks for the food to pass along the tract, which enables the wombat to extract as much water as possible from the faeces. It has the driest faeces of any mammal! The fact that flies live on this food source is another example of how hardy they are.

David McAlpine, an Australian dipterist, has studied these flies more than anyone. He works and collects many flies from around Australia and has done so for a while. One of the flies he collected was so odd that it has never been formally named. It was discovered in the 1970s in southeastern Australia and has been the object of informal debate and deliberation in papers for decades as McAlpine's Fly (in honour of the collector). Although this species has never been described or given a formal scientific name, scientists have worked out that taxonomically it belongs in the Oestroidea superfamily, probably a sister group to the New Zealand bat fly, and that the larvae at least initially feed on dung. This may be the only case of a species that we actually know a bit about still remaining nameless. The fact that no specialist has announced a family to put it in, e.g. blow flies or flesh flies or created a new family, is the real mystery of this fly.

Another group of dung flies and one of my favourites is the black scavenger fly from the family Sepsidae. There are relatively few species globally, about 300, but they can occur in high densities on a delectable pat. I have spent hours watching them wing wave on a faecal dancefloor – a move the males perform to attract the females. This family of flies goes to all sorts of lengths to ensure successful copulation. There are spots on the wings to catch the ladies' eyes whilst the wings are being waggled. The females release odours that are meant to have an aphrodisiac effect, and the males have developed special organs or claspers to stroke, manipulate and hold their mates in place during copulation. I once went to a conference where one of the keynote speeches was all about the courting and reproductive strategies of these flies. During this we watched videos filmed in a research laboratory of Sepsidae stroking, tapping, fondling and kissing (that was a bit much) and it wasn't even 9 am. A wonderful paper entitled *Bending for love* by Nalini Punsamoorthy, a researcher who is also interested in sexual selection, looked into the relationship between how these flies mount their partners and whether this is affected by having modified legs. The legs of the males of many species have hooks or hairs to enable them to successfully mount the females. Nalini and her colleagues found that these structures had evolved just once and all the flies that had them mounted in the same way – the male jumps or climbs onto the female's back and quickly clasps onto the base of her wings to support himself. Those without these adaptations have different strategies. My favourite species in this regard is *Orygma luctuosum* – the amorous male jumps on to the female's back and wraps his first and second pairs of legs around her thorax. He really has to hold on tight initially as she will try to shake him off or roll over, hoping to dislodge him.

Talking of strange sexual practices, the most notorious of the dung flies are from the family Scathophagidae and some of them are just deviants! These are a relatively small family of flies, just 250 or so have been described so far, and they are mainly found in the northern parts of the world. They are commonly called dung flies although this does not apply to all of them. However, in the genus *Sarcophaga* they are dung lovers, and this genus contains my deviants. The females are attracted to lovely fresh pats that are perfect larval habitats, either for feeding on or for providing prey in the form of other dung-feeding larvae. The males hang around the edges of the dung pat, on the surrounding vegetation or more commonly on the dung itself, waiting for a female to come along. And when one poor helpless female does, they launch themselves on her. She is often seen struggling under a pile of males, who are desperately fighting each other to get through to her. The poor little lady is often mutilated in the melee and may lose wings and legs and sometimes her life.

We should be grateful there are so many families of flies that feed on dung and among them are a group of chunky little flies called the Sphaeroceridae, the lesser dung flies. The British entomologist Harold Oldroyd described these flies as being 'inconspicuously successful' as they are everywhere. Both the male and female adults have a large first tarsal segment on their hind legs that is thicker and shorter than the second, and they are generally dark, small and robust little creatures. Many species of this family have taken a novel approach to reaching their food source and their mates, and were first documented 'riding' around on the backs of dung beetles more than 100 years ago. Female flies ride on the back of these beetles to ensure they are at the freshest primary food source for their larvae, whilst the males ride around waiting for the beetles to congregate next to dung to give them a prime spot so as to be the first to see the arriving females. They live in a groove on the beetle's back and are known to stay close to their transport even when laying eggs. A lovely side effect of being on such hosts is that in an emergency they have an extra food supply in the form of anal secretions.

This is not the only fly to ride its invertebrate faeces provider. Another lesser dung fly species, *Acuminiseta pallidicornis*, has been seen riding around on the back of giant millipedes in western Cameroon and specializes in laying its eggs on the frass (millipede waste). Not all are phoretic – physically transported by their host – but they are definitely some of the more adventurous flies. There

Lesser dung flies riding around on the back of dung beetles.

is even a wingless lesser dung fly that hangs out in seabird colonies on several sub-Antarctic islands. One species, *Anatalarta*, develops over a much longer time (five months), and adults can take up to nine months as development is inhibited by the low temperatures. And also to help them survive they are little fatties – the adults may comprise a hefty 40% dry body mass of lipids.

I can't end this chapter without mentioning the hover flies and the infamous rat-tailed maggot that loves nothing more than a pit of slurry. The rat-tailed maggot is the larva of the drone fly, *Eristalis tenax*, and I briefly discussed it in Chapter 2. Its preference for foul sediments or rotten liquids means that it has long been

The larva of the drone fly, *Eristalis tenax*, with the longest spiracle or breathing tube of any fly. The spiracle is telescopic.

associated with humans. The larvae are important indicators of polluted waters similar to those inhabited by the bloodworms, the bright red aquatic larvae of the non-biting midges. This species has a long siphon that is three segmented and telescopic, and can extend nearly six times their own body length. Rat-tailed maggots can reach up to 2 cm (¾ in) and siphons have been recorded at 15 cm (6 in) – the longest breathing tubes of all the flies! These are commonly found in garden waste and at the Museum we often have to answer queries from the concerned public about these alien-looking creatures. Most people cannot believe that anything beneficial can come out of living in such fetid environments and are amazed at how attractive and useful the resulting adult flies are. Without munching adults and maggots, we would be living in a very different environment. We know from studies in the UK that 201 species have been recorded from cowpats and 183 from human waste. We know that most of these cow-visiting flies are consuming the pats, but we know very little about whether those flies on human waste are feeding on it, laying eggs in it or predating on other inhabitants. I'm not advocating we start a citizen science project on this, but once again it illustrates how much there is still to learn.

CHAPTER 5

The necrophages

A friend will help you move, a good friend
will help you move a body.

Steven J. Daniels

A CCORDING TO BENJAMIN FRANKLIN nothing can be said to be certain except death and taxes. Everything living eventually dies. But what happens to all of the bodies? Humans are generally either cremated or, as with most animals, left to rot away. Can you imagine if this didn't happen? Corpses littering the high streets and countryside! Luckily for us, nature's own gravediggers start working on the dead immediately after life ceases. Adult flies locate the bodies, lay their eggs and, for the most part, leave their offspring the job of cleaning up the mess. If it weren't for these little organic munchers we would be knee-high in dead bodies, with corpses floating around in a quagmire of putrefying organs! These necrophagous species, along with the dung beetles, did not evolve until relatively recently – 66 to 26 million years ago – and since they were not around during the age of the dinosaurs, the decomposing fauna and landscape must have looked quite different then. Flies weren't the first to be involved in the clean up then but they are some of the most important insect carrion feeders today in terms of both densities and efficiency. Not only do

The goggle-eyed looking head of a bluebottle larva, complete with fangs.

they munch away on dead bodies but some feed on the dead or dying parts of living ones, which can have a positive or negative impact depending on the situation. The negative is myiasis – a parasitic infection by any fly larvae in a vertebrate host. This often looks as gross as it sounds and I will cover it later. But the positive is a very exciting thing indeed – and one that humans have used for thousands of years to aid our recovery from injuries.

Not too often can one quote the *King James Bible* in a scientific book, but the Book of Job (24:20) states: 'The womb shall forget him; the worm shall feed sweetly on him'. It was not worms but fly maggots that were being discussed here, and this was not the first published mention of maggots feeding on meat. In the 4th century BC, Aristotle, the famous Greek philosopher, proposed his spontaneous generation theory. He argued that some animals grew from their parents while others just appeared – including flies – from either rotten vegetable matter or from the 'inside of animals out of the secretions of their several organs'. This theory was believed to be true from Aristotle's time right up to the 17th century. It was then that Francesco Redi, an Italian physician and nature lover, conducted the first experiments debunking the myth, and these were reinforced in 1862 when chemist and microbiologist, Louis Pasteur, conducted a simple experiment by boiling meat in a conical flask. When the flasks were sealed nothing grew on the meat but when they were left exposed flies appeared. But the flies' appearance was not a miracle of spontaneous generation. They needed to arrive in the first place and lay their eggs on the food source.

You can understand why Aristotle was confused – where did the flies come from and, more pertinent, how did they get there so quickly? Well, we know that some species of blow flies in the family Calliphoridae are able to detect dead bodies from considerable distances – some have been shown to migrate to corpses from more than 16 km (10 miles) away. Dr Madison Lee Goff, a forensic entomologist working in Hawaii, has also found that they can get there very quickly as well – in some cases taking just 10 minutes. And they have to, because a decaying body is, by its very nature, an ever-changing food source and there will be lots of competition for it from different species at different stages.

There are five main necrophagous (body-feeding) families of flies: the blow flies, house flies, flesh flies, soldier flies and scuttle flies, but there are species from many other families that also turn up to feast. The Calliphoridae are arguably the most important in terms of the numbers involved in the decomposition of flesh. Although the family are commonly called blow flies, they also include the bluebottles or greenbottles, the adults of which are some of the most recognizable flies. They are the 'vomiters' and some of the most despised insects alongside house flies and mosquitoes. They arrive at the scene and get to work straight away, laying thousands of eggs on the body, which, within hours in some cases, develop into larvae. And if you thought sewage was a good source of nutrients, imagine how good decomposing flesh is. Once more it is the larval stage that is the important part of the life cycle as the larvae are the consumers of flesh.

There is a predictable succession of flies that arrive at a corpse, with different species of fly specializing in eating different parts of the body at different stages of decomposition. We have used this to help us solve crimes. Forensic entomology is a very popular subject today thanks to shows like *Crime Scene Investigate* (CSI) – in fact the increased popularity of forensic courses is referred to as the CSI effect. But this is not new. Blow flies have been helping us poor-smelling species for a long time now. A classic story, often told but worth retelling, is about the first recorded case of forensic entomology in 1235, in a small village in China. A Chinese lawyer-cum-death investigator called Sòng Cí (or Sung Tz'u) wrote up the 'case' in the medico-legal text book *The Washing Away of Wrongs*.

One of the farmers from the village had been brutally killed with a hand sickle, and Sòng Cí realized there was a cunning way to find the murderer. He asked for all the farmers to attend a meeting and to bring their sickles with them. He then had them wait. They did not have to wait long as the days were warm and flies soon started appearing. And these flies all went for the same sickle. Sòng Cí confronted its owner, who was so shocked that he confessed immediately. What the killer had not known was that although he had cleaned the weapon of all visible signs of blood, the flies would still be attracted by the minute traces remaining on the weapon.

It was another 800 years before flies were used once again to legally incriminate a murderer, this time in the UK. At the Museum we have a rather famous jar of maggots that was a sample from the first criminal case where maggots helped convict a murderer. The story begins with Dr Buck Ruxton, who was a practising GP in Lancashire, England, during the 1930s. He was generally well-liked and respected in the community where he lived and worked.

Then, in September 1935, two mutilated female bodies were discovered wrapped in newspaper in a small ravine in Dumfriesshire, Scotland, 100 miles north of where he lived. Though the newspaper used was a national one, one of the pages used was from a supplement only available in an area very local to where Dr Ruxton lived. And when it emerged that Dr Ruxton's wife and maid had disappeared suspicion inevitably fell on him. He denied any involvement, claiming that the maid had fallen pregnant and that his wife had gone away with her to assist with an abortion. But two key pieces of forensic evidence found him out. First, a comparison of images of one of the skulls with photographs of Dr Ruxton's wife pointed to one of the bodies being hers (a discipline called craniology). Second, a sample of maggots collected when the bodies where discovered was sent to Dr Alexander Mearns, an entomologist at the University of Edinburgh. He identified them as

The most incriminating of maggots, *Calliphora vicina*, a very common carrion blow fly – the larvae that helped solve the Ruxton murders in the 1930s.

Calliphora vicina, a very common carrion blow fly, and he was able to establish that they were somewhere between 12 and 14 days old, which meant that the bodies had to have been there at least two weeks. This provided vital information as to when the murders took place, coinciding as it did with when Dr Ruxton claimed his wife and maid had gone away and, coupled with other evidence, it was enough to lead to his conviction and hanging.

Our understanding of forensic entomology has increased greatly over the past 80 years. In the USA there are numerous body farms where donated human bodies in various states of dress are placed in various positions and locations and insect succession is studied. In Australia the first body farm has just been set up in the Blue Mountains near Sydney and as yet this is the only other facility not in the USA. In the UK we can only work with decomposing pigs and at the Museum we have our very own maggot man (forensic entomologist) Dr Martin Hall, who, with the assistance of many, is researching various aspects of forensic entomology. Up in one of our famous towers that never fail to impress the visitors as they approach the Museum, the entomologists often have rows of Petri dishes crawling with maggots, all being subjected to different experiments. In days gone by, when the entomologists were in a temporary home near the towers, the occasional fly would escape the laboratory and come down the many flights of stairs to join us for lunch. How did we know that they were from upstairs? They were always the most pristine of specimens, iridescent in our dim common room light, as they had been nurtured in the most perfect of conditions.

Body farms are a perfect way for us to study the various factors affecting decomposition – in the outdoors, out of the way of curious humans and hungry vertebrates – and enable us to understand how long different stages take. A body that has been hung, for example, decomposes differently to one placed in a bin. I once attended a conference that discussed many of these different results in detail, the example of a suspended corpse sticking in my mind. Instead of decomposing the suspended corpse became leathery almost tanned, with very little maggot infestation even after a lengthy period of time, as the maggots were not able to attach to and infest it. I talked about this subsequently on a radio show, and shortly afterwards received a charming letter from a very fit (she told me) 80-year-old offering me her body, once she was deceased, for research! Very kind, but I had to decline the offer.

As well as position the succession of maggot feasters can be affected by many external factors, such as temperature and other environmental factors be they local or extremely local. One such modifier is narcotics. Back in the late 1980s, Dr Madison Lee Goff, the Hawaiian forensic entomologist, received a phone call from another entomologist who had maggots from a female who had been stabbed to death. Oddly the maggots were of different sizes and so the implication was that they were of different ages – making the time of death difficult to determine. Now Goff was, at that time, investigating the effect of drugs, specifically cocaine, on maggot development. In his book *A Fly for the Prosecution* he amusingly retells how he had to apply to the Animal Care and Use Committee at his work for permission to give cocaine to rabbits and then go about trying to legally purchase the product! He eventually did get permission but then had to rely on donations from police agencies rather than buying it himself. When he received this call he immediately thought about cocaine but, he admitted later, it was a long shot to ask if the victim had been tested for drugs. Cocaine is a strong stimulant which in humans can initially produce a feeling of euphoria as it mimics adrenaline. Goff found that maggots subjected to cocaine grow more rapidly in comparison to their clean-living companions because of this stimulant. On investigation there were indeed traces of cocaine in the body and the larger maggots were found specifically around the nasal region (snorting this drug is the most common method of ingestion). Thanks to his new research Goff was able to determine post-mortem interval, which in laymen's terms is the time since death, as he was able to work out timings using both the development of maggots under the influence of cocaine and those developing naturally. From this they were able to establish when the victim had died, which had previously been confusing. This new timing did involve her consuming cocaine and was consistent with the other non-insect evidence and so linked her killer's activities to hers.

Forensic entomology is not the only good thing that maggots do for us. As well as helping us determine the time of death, blow flies actually help prevent death in some cases. Species from this family specialize in consuming necrotic parts resulting from an injury or wound as well as decomposing bodies. Open wounds often have necrotic tissue in and around them – commonly called gangrene. If left untreated, gangrene can result in loss of limbs and, if severe, even death.

An ancient but nonetheless effective way of dealing with gangrene was to place live maggots in and around the wound. Legend has it that Genghis Khan always had a caravan full of maggots with his army to enable his soldiers to receive treatment quickly. Moving from a great warrior to a great war, the American Civil War was particularly brutal, with estimates of 750,000 deaths, not just from war wounds but also from many arthropod-borne diseases and poor sanitation. Medical doctors soon noticed though that wounds infested with maggots healed better than those without them and so began adding more to the wounds. In his book *Flies and Disease II. Biology and Disease Transmission*, Bernard Greenberg quotes Doctor John Forney Zacharias: 'During my service in the hospital at Danville, Virginia, I first used maggots to remove the decayed tissue in hospital gangrene and with eminent satisfaction. In a single day, they would clean a wound much better than any agents we had at our command. I used them afterwards at various places. I am sure I saved many lives by their use, escaped septicaemia, and had rapid recoveries.'

There are many other examples, both historical and current, of maggots removing rotten tissue, technically called debridement, and nowadays there are factories globally dedicated to the production of debridement flies. The most common blow fly used is *Lucilia sericata*, the common greenbottle. Research has shown that the maggots of these flies secrete allantoin, an antiseptic compound, the production of which is not just restricted to these flies but is across the animal kingdom and in some plants. This magic compound stimulates healing and cell development and is now used in many products including cosmetics. It has also now been determined that the effective molecule in allantoin is urea, a major component of human urine, a compound which not only stimulates cell regeneration but also keeps infections at bay.

The ability of maggots to release 'healing fumes' has also been shown to be effective in reducing TB and MRSA infections. Around 100 years ago Arthur Bryant, a Yorkshire man who bred maggots for fishing bait, realized that they could be helpful in other ways and set up 'maggotoriums'. If you have a sensitive stomach I will almost apologize for this next bit of the story. Each summer he received on average 18 tonnes of dead animals (usually from zoos) and he would leave these in the woods to attract greenbottle maggots (the smell was apparently detectable three miles away). Once maggots were visible on the meat, it would be removed and transferred to these maggotoriums. Stephen Thomas writes in his book *Surgical Dressings and Wound Management*: 'Consumptives would sit in the maggotorium beside troughs of maggots to inhale the fumes and pass the time reading, chatting or playing card games'. However, mass production of penicillin and sulphonamides and, I am hazarding a guess, the smell, ensured that these maggot factories never really took off. Now that there is widespread resistance to antibiotics, scientists are once more researching the antibiotic properties of maggots, determining the important components of these maggot fumes and finding a way to apply these to bacterial infections that are developing resistance, such as MRSA.

Apart from having this chemical benefit, maggots are just very good at cleaning up dead flesh. Once properly sterilized, they can eat away at the dead or dying flesh in wounds to leave them clean, thus enabling tissue to regenerate. Suppurating diabetic ulcers and other rather grim infections or injuries can be stuffed with sterile maggots and left for a couple of days before the maggot dressings are changed or removed. They do not eat undamaged cells and, even if left in the wound – though not exactly recommended –

they do not cause any damage to the patient. Ronald Sherman MD wrote in 2003 about the efficiency of maggots versus conventional therapy: 'Non-healing diabetic foot ulcers account for 25–50% of all diabetic hospital admissions and most of the 60,000–70,000 yearly amputations in the US.' But Sherman found there was a significant reduction of necrotic tissue and amputations when maggots were used instead of standard methods.

You don't need to pour maggots into a wound any more either. Now doctors can order what look like teabags – albeit teabags stuffed with maggots – instead. These maggot-bags enable the maggots to poke their little jaws through the mesh and get at the necrotic tissue, while the mesh stops them wandering off. And, as we already know, their waste products are useful to our wounds. The bags get

The adults of the common blow fly or greenbottle, *Lucilia sericata*, which secrete an antiseptic compound that stimulates healing.

changed every couple of days, and after five or six weeks you have a beautifully healed wound.

Maggots don't just devour the meat course but also favour the cheese course. There are a group of flies called cheese skippers or Piophilidae, and they all have a liking for animal products, including mouldy cheese and cured meats. A particular type of traditional Sardinian cheese called *casu marzu,* is banned in many countries because it contains live *Piophila casei* maggots, commonly called cheese skippers. The feeding activity of these maggots causes the insides of the cheese to decompose and then ferment, resulting in a soft, liquid centre. You can either leave the maggots or choose to consume them with the cheese as part of the experience. Whether you choose to eat them or not, caution is advisable as, when startled, the maggots can jump distances of up to 15 cm (6 in). They do this by raising their anterior end, bending it forwards in a loop until their mouth hooks attach to the end of their abdomen. When the maggots rapidly straighten their bodies, it causes them to 'jump'.

I have not eaten this cheese, but was prepared to do so in the name of science (and curiosity) until I did a little more research into this delicacy. This species of fly, and similar species, are extraordinarily resilient. They can survive in our digestive tract and its acidic juices, and can often lacerate the gut lining with their mouth hooks. Internal bleeding and stomach spasms, brought about by a cheese, does not sound like a good ending to a meal. Accidental myiasis resulting in diarrhoea, pain, nausea and other gastric symptoms is apparently not that uncommon but something that I would rather avoid given the choice. There have even been cases reported when larvae have pupated inside the body and an adult fly has emerged along with the faeces.

Most adult cheese skippers have rather oddly shaped heads – almost cheese-wedge-shaped. The males of another in the group, the waltzing fly, *Prochyliza xanthostoma*, have quite wonderful

The male waltzing fly, *Prochyliza xanthostoma*, with its conical shaped head – not only does it look good for the female but it will fight for her too.

heads – almost conical in shape with very large and thick antennae – and they feed on animal carcasses. As their name suggests the males dance to woo the females. But they also have to defend their territories (the carcasses) from other males and can have the most amazing fist fights – they really reign down punches on each other.

Yet another species of Piophilidae is the antler fly, *Protopiophila litigata*. These don't have antlers and should not be confused with the antler flies in the fruit fly family that do. Instead *Protopiophila litigata* are found on either antlers that have been shed or on the antlers of dead deer and similar species. Their scientific name hints at their behaviour – litigious or aggressive. As with the waltzing fly, males can get very physical with each other but they can also cluster together in groups. This conflicting behaviour is the result of the considerable amount of mate-choosing by both the females and, more unusually, the males. The male is looking for a female with a large derrière, holding out for a lovely big one as this indicates that her eggs are very close to maturity and so the chance of him impregnating her is at its greatest. She is looking also for a large dominant male as it's presumed he will be better at protecting her and will have more ejaculate (which she partly expels then ingests after mating). Romance is alive and blooming with these flies.

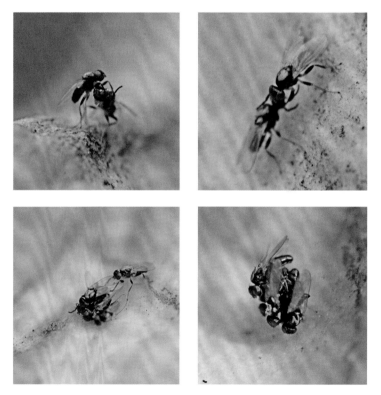

These male antler flies, *Protopiophila litigata* – found on antlers as opposed to having antlers – fight with each other and in groups for one reason only, to win a mate.

Two evolutionary biologists, Russell Bonduriansky and Ronald Brooks, describe the mating behaviour of antler flies as males leaping on to 'the backs of females, briefly tapping the sides of her abdomen with his legs, then stimulating her abdominal top with his tarsi, parameres and gonopods [specialized external reproductive organs] until she extends her genitalia, after which the pair established genital lock'. These two researchers watched these flies having sex for periods up to 10 minutes, to work out the different behaviours and timings. And this wasn't confined just to the laboratory, they also studied the flies outside in their natural environment on antlers.

I have swept flies from this piophilid group away from rotting cow skulls while collecting in Peru, and all that was left were tiny portions of tanned skin stuck to the bones. Harold Oldroyd also wrote about these flies in *Natural History of Flies* in connection with a collection of 17 elephant skulls left outside the window at the Museum to decompose in the fresh London air. Scientists have been using insects for a long time to help clean skeletons of soft tissue, whether for research or display. Nowadays most skeletons aren't left lying around outside but are kept in special enclosures called dermestariums, within which dermestid beetles carry out the cleaning up of the specimens. Oldroyd comments that 'a fine collection of *Piophila* was a small compensation for the smell...' I am curious to know what the public made of these elephant skulls decomposing in the sun. The Museum has a dermestarium but it is only used for smaller animals.

Another group of flies that have a liking for dead bodies are the Phoridae, or scuttle flies. These have incredibly diverse feeding preferences but one group in this family is referred to as the coffin flies because, indeed, they develop inside corpses. They are able to survive in these unique environments for many generations even when the corpse is buried in a coffin. The most important of these flies is *Conicera tibialis,* which has a global distribution. The larvae start munching away once the corpse has dried out, which can be as soon as a year after death, but specimens have been recorded at post-mortem intervals of three to five years. Daniel Martin-Vega, of the forensic research group at the Museum, and colleagues found fresh adult flies on a corpse exhumed 18 years after death. When we look at the succession of species on a corpse, flies will be some of the first animals to arrive and some of the last to leave. For a fly to get into a coffin buried several feet under is no mean feat, but these flies are never more than 3 mm (⅛ in) in size. Adult *Conicera tibialis* have been timed descending from the surface and they were able to burrow to a depth of 50 cm (20 in) in about four days. This species has wings, but many other species of coffin fly are apterous, meaning wingless, such as *Puliciphora borinquenensis.*

Many species of coffin fly, such as this female *Puliciphora* sp., are wingless and burrow down to get to the corpses.

Puliciphora borinquenensis have not only an interesting diet and female morphology but also several different mating rituals. The female starts courtship with vigorous abdominal pumping which is thought to waft her pheromones around to attract males. She parades around near the original oviposition site from which she emerged as an adult. Males then decide to adopt one of four different reproductive routines. The first, and in my mind the most ambitious, is for the males to remain stationary and grab as many of the parading females in succession as possible to mate with over about 30 minutes. Peter Miller, an entomologist at Oxford University, stated that they mate with 0.66 females per minute over this period. He observed that one particularly successful male performed 45 separate copulations in half an hour. The second option, again with the males just standing still, is to wait at the oviposition sites and grab the non-parading females that had either already copulated or were not yet ready to. The third and fourth options are more amusing, as the parading females are 'airlifted' by the males, while carnally attached, to new egg-laying sites when these can be found, or dumped randomly when they can't! And just when you thought their behaviour couldn't get any more extraordinary, if the males that are carrying females do figure out a good oviposition site they can transport large numbers of females there – as many as 30 at very quick intervals, that's one female every two minutes. What is there not to admire about the behavioural adaptations of flies? The density of individuals and the age of the males determine which of the four reproductive pathways they choose.

Scuttle flies don't just like human corpses. The larvae of several of the *Megaselia* genus (a rich genus of more than 1,400 species that contains nearly half of the described species from this family), have been found feeding on dead snails. Some of these species were initially considered parasitic as the female was seen laying eggs on live insects. But no, the larvae of *Megaselia scalaris* are in fact saprophagous, meaning they eat decaying flesh, and we have determined this by their mouthparts. This species, as with many saprophagous larvae, have pharyngeal ridges in their mouthparts. These ridges sieve out the ingested fluid – the decomposing bodily juices – and enable the larvae to ingest solid rotten patties rather than litres of putrefied juices. Efficient, if a little nauseating. Amazingly, the mothers are able to determine whether the prey is injured or not (we don't know how), and thus select a suitable host for their larvae.

Scuttle flies like both fresh and seasoned flesh and this is not uncommon across many families of flies that have necrophagous species. House flies and moth flies are often found grazing on mature and decomposed corpses and so house flies are often some of the first and last to arrive at a body. Forensic investigators need to be very careful about the accuracy of identifications as wrongly identified species may lead to incorrect ageing. The number of families that have been recorded turning up on carcasses is vast. Work by Catarina Prado e Castro, a Portuguese forensic entomologist, with decomposing pigs has highlighted this as she and colleagues identified 43 different families of diptera on corpses.

Whether it is getting rid of the bodies or mending them, diptera are essential in our environment. These flies favour environments that most of us would consider highly unfavourable, and convert the defunct to functional, which makes for a more pleasant world for us all.

CHAPTER 6

The vegetarians

I am not a vegetarian because I love animals;
I am a vegetarian because I hate plants.

A. Whitney Brown

CATERPILLARS OF MOTHS, butterflies and sawflies are commonly known for being some of the greatest scourges to gardeners because they eat their vegetables. Aphids are equally demonised for sucking the life essence out of plants – especially my mother's roses. But we tend to ignore the voracious vegetarian flies. And the impact of these vegetarian flies on our populations can be as damaging to us as the flies that carry human diseases. Don't get me wrong, flies don't roam around in majestic hoards across vegetable plots eating everything in their path like the cabbage white butterfly does, or feed on crops in vast swarms like locusts do (a recent swarm in Argentina, started in 2015, reached over 6 km (3¾ miles) long and 3 km (1¾ miles) high. But they are nevertheless an important part of the ecosystem).

There are 40 families of flies that contain herbivorous species, but few of these families are dominated by them. All herbivores, be

A gall midge, *Fergusonina* sp. which lives in galls in myrtle that nematodes living in the fly induce the plant to make – the only known example of symbiosis between insects and nematodes.

they insect or other animal, eat the primary photosynthesizers – living plants, algae or bacteria. These photosynthesizers are not as nutrient-rich a food source as decomposing plant material, which can be laced with microbes, or putrefying flesh from animal bones, but they do provide a very important element, nitrogen, which is an essential component of amino acids, the building blocks of all proteins, and of urea, equally essential for ensuring the removal of toxic waste products.

Sadly for the vegetarians, the nitrogen content in plants is quite low in comparison to animal tissue. Although air is composed mostly of nitrogen (78%), the nitrogen is not accessible to many plants in this form. Nitrogen in the air is composed of two atoms stuck together, which are very difficult to prise apart. Separating these two atoms to make the nitrogen element usable to plants, a process called fixing, does occur naturally when the atmosphere is struck by lightning, but this is hardly a reliable event. So, plants have evolved two further, more reliable methods to ensure they obtain enough nitrogen. The first is to take up nitrates that are in the soil (or in fertilizers), and the second is to absorb the soil ammonia that has been fixed either by free-living or symbiotic bacteria. In whatever way plants obtain this essential nutrient, they will never be as nitrogen-rich as animals – animals average 50% of body mass as protein while in plants this is only 2–5% of their make up.

And why is this relevant? Well, irrespective of their diet, flies' nutritional needs are similar across the entire order. To compensate for the lack of proteins in their diet, many vegetarian flies have either extended their larval stage – the primary feeding stage – or have concentrated their feeding activities on parts of the plant that are nitrogen-rich such as the roots, where many of the plant's nitrogen-fixing bacteria are, or the fruit bodies.

Terrestrial fly larvae evolved and adapted to feeding on plants relatively recently, within the past 100 million years. The earliest species of fly had aquatic larvae; some of these were herbivorous (feeding on algae), but most were saprophagous or predators. All adults are suctorial feeders – technically many are herbivorous but they feed on plant products such as pollen and nectar, rather than the plant itself and it is these that this chapter focuses on.

When it comes to immature flies you rarely see them munching on a leaf like caterpillars because, in many cases, they've chosen to go underground, feeding on the roots and tubers in the soil. Some, we could say, have gone undercover as they have taken to living inside the plant, sometimes just below the surface but in other instances in elaborate internal structures. They are entomological spelunkers in a botanical cave. Their presence can create botanical houses in the form of galls in a huge range of plant species – essentially these clever creatures get the plants to grow little houses for them.

As well as gaining food from the plant, the larvae also gain protection, either from natural enemies or the environment. Dipteran larvae are not as tough or robust as their beetle brethren, and they don't have toxins and spines like many of their caterpillar cousins and so often have to hide to protect themselves from their enemies. This method is not totally effective though as there are enormous numbers of parasites that still manage to feed on them. Within the order Hymenoptera, which includes bees, wasps and ants, there are many parasitoid groups specializing in attacking the poor defenceless flies. The larvae of the holly leaf miner fly, *Phytomyza ilicis*, in the leaf miner family Agromyzidae, leaves distinct trails called mines as it burrows under the surface of holly leaves. This one species alone has at least three species of parasitic wasp feeding on, or rather, in it as well as being preyed on by birds such as blue tits.

Not all plant-feeding larvae have to face so many threats and some do enjoy protection offered by their environment. They are sheltered from excess ultra-violet, wind and rain – though all of these advantages disappear if the leaf they're in falls to the ground.

Diglyphus isaea, a parasitic wasp that lays eggs inside the larvae of the holly leaf miner fly. The larvae burrow under the surface of holly leaves, ostensibly to escape such predators.

It's not easy for them to crawl up into a new leaf, and generally these larvae are small, so very small – imagine living between the tissues of a leaf membrane, you can't exactly let yourself go and pile on the pounds. It's also time-consuming finding a new leaf and therefore expensive on their bodies' resources. There are a few exceptions to this and those are mostly with the wood-boring species.

Within nine families of diptera, some larvae have completely adapted to living inside plants, and these adaptations have evolved independently at least six times across the order, resulting in flies known as miners, borers and gallers. Examples of these are the gall midges in the family Cecidomyiidae, which include both galling and wood boring larvae, and Agromyzidae, which include both miners and gallers. The most primitive of the mining families are the limoniid crane flies of the Limoniidae family. Of the 10,500 described species only one mining species has been described, and it lives in Hawaii. It was originally described by an English entomologist, Percy Grimshaw, in 1901 as *Dicranomyia kauaiensis* from six adult specimens collected during several British expeditions to the islands. It is now named *Dicranomyia (Dicranomyia) kauaiensis kauaiensis.* Even for me that is a particularly silly name (genus, sub genus, species and subspecies) and not one that rolls easily off the tongue, but the entomologist wanted to make it absolutely clear that it was the true original genus in terms of morphology and species. It was then described again by a different author who hadn't realized it already had a name, and this new name subsequently became the junior synonym, i.e. when a new name is applied to a species incorrectly as the original name is still valid. This new describer, American entomologist Otto Swezey, visited the islands in 1913 to search for flies. He thought, incorrectly as it turns out, that he had found a new species, which he called *Dicranomyia foliocuniculator* but it was indeed Grimshaw's species. This may have been because Grimshaw had only described adults whilst Swezey found leaf mines (the tracks and homes of the larvae) from which he was able to successfully rear some adults. Swezey's adults didn't fit the earlier descriptions of Grimshaw's species, and Grimshaw had made no drawings of his adults. This, coupled with no descriptions of the mines or the larvae, did not help Swezey with comparing his specimens with Grimshaw's (to be fair to Grimshaw he was only handed adults). Taxonomists are not always clear in their descriptions and these little mishaps often happen. Many species have been described multiple times – some species of mosquitoes have more than 30 synonyms. And

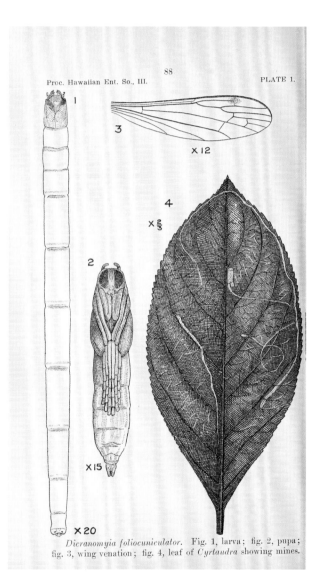

Proc. Hawaiian Ent. So., III.

88

PLATE 1.

1

3

×12

4

×⅔

2

×15

×20

Dicranomyia foliocuniculator. Fig. 1, larva; fig. 2, pupa;
fig. 3, wing venation; fig. 4, leaf of *Cyrtaudra* showing mines.

Original published diagrams of the leaf mining crane fly,
Dicranomyia kauaiensis kauaiensis.

although he was wrong about it being new to science, Swezey was the first to describe the mine, and his paper is still important as it is the original description of that life stage. It's a pity it wasn't a new species as scientists rarely get to claim they have doubled the numbers of described species within a specialized feeding group.

The long-palped crane flies within the closely-related Tipulidae family also include herbivorous species, but these are grazers rather than miners. Across the UK, a very common crane fly, *Tipula paludosa*, is a pest in grasslands and spring cereal crops, whilst a more northerly distributed species, *Tipula oleracea,* is a pest of the winter cereals, especially if the cereals were planted in a field that had previously had a crop of oilseed rape. Commonly called leatherjackets, these larvae generally live just below the surface of the soil munching away on the cereal roots. If it's been particularly wet they emerge from the soil, to consume the exposed lower parts of the plant. Back in 1935 this resulted in the most English of problems when these munching monsters were so numerous they stopped play at Lord's Cricket Ground, in London! Several thousand larvae had to be removed by ground staff and burned as they caused 'bald patches on the wicket', resulting in unpredictable conditions for spin bowlers for most of the remainder of the season.

Crane fly larvae need to consume a lot of food because, more often than not, the adults don't feed at all. Most people confuse this family with mosquitoes or anything that stings, as females have a very pointed ovipositor. They are not dangerous bloodsuckers though and often have very primitive mouthparts that would be hopeless at piercing skin. A classic example of this mix-up features prominently in the 1993 film *Jurassic Park*. Many a performance must have been disrupted due to excessive tutting by indignant dipterists as the onscreen scientists attempted to remove 'dino DNA' from the blood inside the abdomen of what they thought was a female mosquito but was clearly a crane fly. The first shot is indeed of a mosquito – but it's a species where both the adult male and the female are vegetarians. But the long extraction shot is of an adult crane fly that doesn't even feed, let alone feed on blood. And, as a final insult to movie-going dipterists, the crane fly was a male! It has massively enlarged genitalia totally unlike the lancet-like female ovipositor. And be careful fans of the film if you are thinking of purchasing replica amber specimens of mozzies – most of them are crane flies too (although I guess that's authentic to the film).

Although few plant-feeding larvae are detrimental to plants, there are some that cause considerable damage, not just because they are feeding on the roots, but because they have been shown to spread root disease. One example is *Bradysia impatiens* (I imagine them sitting around tapping their feet), a dark-winged fungus gnat from the Sciaridae family, which eats roots and algae from the upper region of the soil. The adult flies transmit a range of soil pathogens that have previously become attached to their bodies and faces. And it's not just the direct feeding of the larvae that is problematic. Further damage comes from the now weakened plants not being so good at fighting off other infections.

The five most economically significant of all the vegetarians are the fruit flies (Tephritidae), gall midges (Cecidomyiidae), vinegar flies (Drosophilidae), leaf miners (Agromyzidae) and rust flies (Psilidae). These are some of the most destructive little feeders on the planet. Arguably caterpillars consume more and as such are much worse, but people tend to forgive them when they metamorphose into charismatic mini-fauna – such short-sightedness. Back to the islands of Hawaii where vinegar flies have gone beserk. There are 550 endemic species, i.e. they are only found in Hawaii – in fact over 90% of all Hawaiian vinegar flies only live there. The islands are isolated from the mainland and so invading species had to adapt very quickly to fill different ecological niches or face fierce competition from existing populations to survive in existing niches. Hawaii has the highest diversity of endemic flies per land unit of anywhere – there are more than 1,100 endemic species of flies there compared to, for example, the British Isles, where we believe there are perhaps only one or two endemic species of fly despite the land mass being twelve times the size of Hawaii. The distance from neighbours really counts.

In their natural environment many species of vinegar fly are perfectly harmless, but there are a number of cases where invasive species have been anything but. *Drosophila suzukii* for instance has been in Hawaii since the 1980s, having spread from Asia, and is now winging its way around Europe. This is of great concern to fruit farmers as this species targets the healthy soft fruits of species such as strawberries. Initially it was thought this would have a massive impact on UK production, but so far, even though it has turned up, we appear to have dodged the bullet in terms of agricultural damage.

Fruit flies have an enormous global economic impact and, as their name suggests, mainly attack fruits, with mangoes and olives being some of their most commercially important targets. And they don't just feed on the fruits of these plants, they also attack the leaves, stems, roots and flowers, and some even form galls. An example of the latter is the goldenrod gall fly, *Eurosta solidaginis*, which is mostly found in North America and which attacks the stems of goldenrod plants, During the winter the northern parts of this continent can get very cold and these little gall-dwelling larvae have to survive temperatures of -40°C (-40°F). The tiny individuals cope by ceasing to eat and grow and, with the assistance of large concentrations of glycerol – nature's antifreeze – in their bodies, they hunker down and wait out the winter.

There are around 5,000 species of fruit fly found globally, and out of these 250 species have been recorded attacking fruits that we would consider of economic importance. In 2007 the World Bank established that Africa as a continent generated over $16 billion in foreign exchange from fruit exports alone and an estimated further $6.5 billion estimated for its domestic markets. But this was nowhere near what could be achieved, as probably more than 50% of the productive volume at the time was being affected by fruit fly infestation.

The Australian government is so concerned about incoming pests that passengers on inbound flights are sprayed with a mild insecticide and all domestic carriage of fruit is banned between states. And it's not just the hot countries that have problems. The UK has many such invasions and outbreaks, often created accidentally through imported plants. In the USA there are not only standard drug-sniffing dogs at the airports but fruit-sniffing ones too. Cocaine or apples, take your pick, but it is equally illegal to import them. The Oriental fruit fly, *Bactrocera dorsalis*, is one such pest and has been shown to be a serious threat to over 150 plant species wherever it has become established. Once more it was poor Hawaii that suffered considerably when it arrived, with more than 125 plant species found to be infected.

At least when there is an attack on fruit the culprits are easy to identify as fruit flies and their relatives have a distinctive abdominal shape, and are generally some of the prettiest-looking flies as they mostly have patterned wings. The females, as with all mothers, want to ensure their offspring develop in a safe environment, but plants and their products try to protect themselves by being quite tough to penetrate. To overcome this, females have evolved a stiffened ovipositor (most diptera have a flexible one) ending in a hard tip that enables them to insert their eggs into the plant tissues. Most people see this hardened ovipositor and panic that the fly is some form of super wasp. However these families are grazers, not fighters.

Of all the fruit flies, the medfly *Ceratitis capitata* is probably the most economically damaging and may be one of the most problematic of all fruit pests. They are polyphagous, that is they eat a wide variety of foodstuffs and attack more than 250 different plant species. This is not the case with all fruit flies. Some are very picky such as the olive fly, *Bactrocera oleae*, which, as you may have suspected, only feeds on olives. The medfly is far more flexible and has successfully managed to travel and infest countries all over the world, helped also by its ability to tolerate wide temperature ranges. In a bid to control it, Los Angeles for example releases millions of sterile male flies every week to suppress outbreaks of imported wild flies (the larvae often come in on imported fruit). This sterile insect technique as it's known is the practice of mass rearing sterile males which, once released, copulate with all the local females – much to

The very attractive but problematic medfly, *Ceratitis capitata*, which attacks more than 250 different plant species.

the chagrin of the resident fertile males. Females then often don't mate again, assuming they've already become impregnated. These reared males have benefited from a plentiful diet unlike their wild cousins, which comes in handy as the male provides the female with food whilst mating. Romantically called accessory gland products, these are essential for sating the female and, together with the presence of sperm, switches off her receptivity to other males. Males that can't feed the ladies don't meet their needs and so lose out.

It's not all bad news with this family. We have started using some fruit flies to help us control invasive plant species. One such plant is *Chromolaena odorata*, an invasive species often called the devil's weed which was originally from the New World tropics but that has been accidently introduced into countries all over the globe. The Centre for Agriculture and Bioscience International (CABI) states that this is considered to be the world's worst weed as it is highly dispersive upon arrival in a new habitat. The gall fly, *Cecidochares connexa* (no common name), from South America, along with a moth, is being trialled as a control species to feast on this beast, with some success in Guam and Micronesia.

Most members of the fruit fly family have very pretty wing patterns, and many have spectacular bands on their eyes. Some of them have taken this wing patterning and run with it. *Goniurellia tridens*, which lives in the Middle East, caused a little bit of a storm a couple of years ago due to the remarkable patterns on its wings. It looked as if they had ants or spiders on them. If this is what the pattern is meant to be then it begs the question: how did this come about and why would they have these patterns? To date we don't know what they are meant to represent, indeed if they are meant to represent anything at all (to me they look like spiders), but they are more than likely used in signalling. Dr Brigitte Howarth, of Zayed University in the United Arab Emirates where they are found, thinks they have probably evolved to confuse predators, both large and

Is it an ant, a spider or a fly? *Goniurellia tridens* with wing patterns that are confusing enough to fool its predators.

small. A spider is less likely to attack if there appears to be another potential predator huddled closely to the prey. And ants also will not approach if there are already ants on the scene.

Many species of fly, including the fruit flies, have wings that may appear quite plain to us but actually may have remarkable light interference patterns caused by light waves reflecting from the upper and lower wing surfaces interfering with each other. If the wings are held against a dark background they reflect a wonderful kaleidoscope of colour. These are not made from pigments, but are in fact caused by the actual size and shape of the wings. They have tiny ridges and folds which, when hit by light, reflect the light back at different wavelengths, causing the changes in colours. These vary across species – so helping us to identify them – and also across sex. That is a nice indication that wing patterns may be used in the subtle art of wooing females!

The most impressive of all stalk-eyed flies, *Achias rothschildi*. When the adult emerges from the pupa it swallows air and pumps it through to inflate its own eyes.

The family Diopsidae – the stalk-eyed flies – are also pretty striking. As their name suggests, many have eyes on stalks, mostly the males, and they use them as indicators of how 'fit' they are to the local females. Despite their good looks, the larvae aren't seen as 'good' to us as they feed on many important crops including rice.

The condition of being broad-headed is termed hypercephalic, and this is found in a further seven families. The females use this feature to determine the male's genetic strength – if he can develop eye stalks that are long and cumbersome and maintain his territories then he must be a worthy mate. Work by entomologists David Grimaldi and Gene Fenster determined that this condition has no common ancestral route but rather has evolved independently in different species or families 21 times across the flies. They also found that all the males that developed stalk eyes defended territories where feeding and ovipositing took place or where there were nocturnal resting places. Those species without stalk eyes tended to exhibit less territorial behaviour.

What is more amazing about the eye stalks is that the flies are not born with them – the larvae don't have crazy eyes. It's in the pupae that the eyes start changing. Initially they develop in a similar way to non-stalk-eyed flies, but gradually the optic nerve extends into a sprung coil and the cuticle that surrounds it becomes corrugated, much like the bendy bit in a bendy straw. At this time it is still soft. Once the adult has emerged from its pupa it then swallows air through its mouth and pumps this through the still-transparent eye stalks to inflate them. These stalks gradually expand to their full capacity and harden. This process is similar across all the stalk-eyed families, although intriguingly sometimes the antennae are located with the eyes at the end of stalks (as in diopsids) whilst in other families they are still on the main part of the head.

Flies, especially the vinegar flies and diopsids, are model species for understanding mating systems due to their evolutionary diversification and ecological variation in both habitat and feeding preferences. In diopsids, males have been observed, when the females appear to become a little bored or restless during the mating process, to start massaging and stroking them. In the genus *Diopsis* it's the other way round, with the females massaging the males – maybe that's her not-so-subtle hint to him to get on with things!

When it comes to mating rituals, the signal flies of the Platystomatidae family have some of the most complex. Not all of the larvae of signal flies are vegetarian but one genus, *Rivellia*, feeds solely on nitrogen-fixing root nodules of many plants including soybeans. Once more it's the adults' mating behaviour that intrigues me about these vegetarians. Females have been seen running around on leaves, sometimes in spirals or circles, wing waving as they go, occasionally pausing for a second or two. The smaller males mimic this, chasing after the female and, when possible, touching her abdomen with their proboscis or front legs. Eventually she allows the male to mount her and, with his copulatory organ extended, he taps her abdomen with it to signal the commencement of copulation. The Australian entomologist David McAlpine studied four species of signal flies including *Euprosopia anostigma* in the 1970s. Males of this species tapped the female's wings or abdomen and also ingested her anal fluids.

It's not just with eye stalks that the males have gone crazy for body modifications. In the fruit fly family you can also find the goat flies and the moose flies – collectively known as the antlered flies. These are so called because they have massive extensions growing out of their heads that resemble those of a deer. And they use them in the same way too! The antler flies are from one genus, *Phytalmia*, and the males of seven species use their antlers to defend oviposition sites. In a similar way to the stalk-eyed flies, the individuals judge each other based on the size of their protrusions.

The biggest family of miners are the Agromyzidae – the leaf miners. I have spent a lot of time trying to locate them up in the mountains of Peru. For most of the 3,000 species that have been described globally, the larvae are borers and drill through the leaves or stems. One of the more serious species in terms of pest status is the bean fly, *Ophiomyia phaseoli*. It launches a two-pronged attack on many tropical legumes, with the adults boring through the stem to lay the eggs, and then the subsequent larvae mine the leaves – a very industrious species.

Another infamous borer is the hessian fly, *Mayetiola destructor,* found in the gall midge family Cecidomyiidae. This little fly is a serious pest on wheat. It was originally from Asia, but was apparently transported to Europe and then on to the USA in the straw bedding carried by Hessian troops from the German state of Hesse-Cassel. These troops were hired by the British during the American War

A female hessian fly, *Mayetiola destructor*, with her exposed ovipositor producing pheromone attractants.

of Independence, and ironically this fly has probably caused more long-term damage than the troops ever did. Female hessian flies select the host plants on which to lay hundreds of eggs by their physical characteristics, such as hairiness. They can produce up to five generations of flies per year, although two is more usual. It's easy to imagine how vast amounts of damage can rapidly happen – a large infestation in 1836 in the USA was thought to be one of the contributory factors for many American farms failing just ahead of the financial panic of 1837 and so adding to the years of economic hardship that were to follow.

Many methods are employed to control these pest species, ranging from chemical to physical and biological. Included in this was research into pheromones – the chemical substances produced by both sexes as a method of communication. Most orchards use pheromone traps that mimic female pheromones, luring males into a container or sticky trap, which can be regularly monitored to determine pest densities. This is a cheaper alternative to regular blanket spraying with insecticide and avoids adding yet more chemicals to the environment, as spraying only occurs when densities reach some critically defined level. Pheromone research is not only useful for monitoring, however. For many years wheat growers had been breeding resistant crops, but in the 1970s flies were finally able to overcome the enhanced defences of the crops. So in the 1980s research began once more into how to prevent the flies decimating crops, including looking at the pheromones of these flies and the factors influencing oviposition. Some of these investigations involved putting flies into wind tunnels and studying how the females' scent affected the males under different environmental conditions. Wind tunnels can also reveal how wind speed affects oviposition – the greater the wind speed the more eggs the female lays, presumably as she is more confined to one spot! Research has also focused on how humidity and light intensities affect her flying (once more in tunnels) and all of these findings aid us in understanding their behaviour and subsequently manipulating it so they are less destructive to our crops.

As well as the borers and the miners, many of the phytophagous flies are gall formers. The galls are abnormal growths, effectively cancers of the plants, and are caused not just by flies but also by wasps, mites, fungi, bacteria and other parasitic invertebrates. Ephraim Putt, a New York State entomologist, described the gall formers as being similar in character to the biblical King Solomon: 'Consider the gall insect – it does not sow yet it reaps; it does not build yet it is sheltered; it gives nothing and receives abundantly'. The gall midges of the Cecidomyiidae family are a notoriously difficult family to identify – both in the larval and adult stages. The galls that are produced though aren't, as they are often highly complex structures and easily identifiable as galls are specific to their host.

Most gall midge identification was originally undertaken by botanists rather than entomologists, and many of the latter would still rather look the other way with this family. As adults they are incredibly difficult to tell apart and so must remain nameless. Hopefully this is where molecular identification techniques such as DNA barcoding or whole genome sequencing will become very useful and we will be able to use these genetic markers to differentiate between the different species.

Galls are perfect little homes for the flies and protect them from many of the ravages of life. Some of them even invite other animals to come and live with them. The family Fergusoninidae is just such an example. With just one genus, *Fergusonina*, these flies are found in Australia and the Orient and have decided to become bedfellows with nematodes. This relationship is only found in myrtle plants and predominantly only in Australia. Evolutionary biologist Leigh Nelson and colleagues found that the presence of the nematodes induces the plant to make the galls. The nematodes live inside the flies and use them to provide food as well as transport. This is the only example known to date of insects and nematodes being co-dependent.

Destructive, beautiful, localized or far-reaching, whoever said that vegetarians were dull has never observed these little gems!

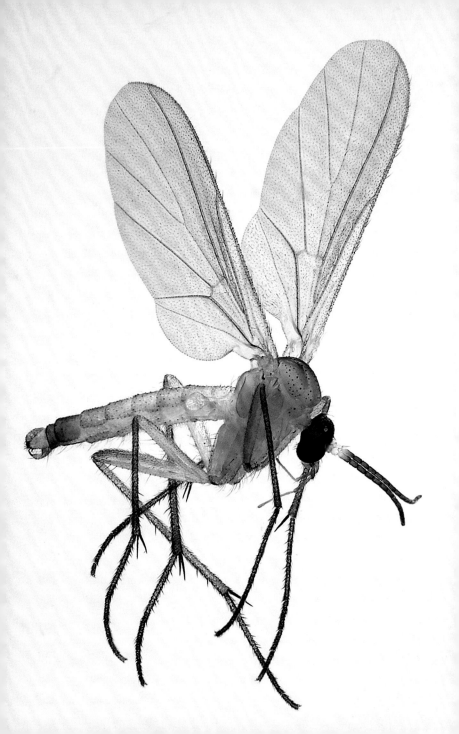

CHAPTER 7

The fungivores

Well, a peach has a lovely taste and so does a mushroom,
but you can't put the two together.

Arthur Golden, Memoirs of a Geisha

THE TERM VEGETARIAN is slightly misused by humans as we use it to describe people who are both herbivores (plant eaters) and fungivores (fungus eaters). In fact fungi, including mushrooms, are in their own separate kingdom and are more genetically related to animals than they are to plants. They are a valuable food source and many species have evolved to feed on them. Many of the flies that have been found feeding on fungi may also be described as saprovorous-feeding species – they feed on the fungi as well as what the fungi itself is feeding on. This chapter however focuses on the truly fungus-feeding flies that feed on the fruiting bodies (sporophores), the spores or the mycelia.

Most fungivorous flies, also called mycophagous flies, consume food only during the larval stage of their life cycle. More than 25 families of flies have some species with mycophagous larvae and two of them, Bolitophilidae and the flat-footed flies Platypezidae, are solely comprised of them. Many of them are shy little species, where the larvae hide away from prying eyes, living between the gills

The fly, *Diadocidia spinosula*, and the name is all we really know about this fly.

of the fruiting bodies. Other species feed on mycelia, the branching thread-like strands that arise from the main fruiting bodies, and can be found free-living in the soil or on decomposing/infected plants.

Although there has been limited research into flies, their life histories and feeding behaviours, there have been some extraordinarily comprehensive studies done on the fungivorous communities, mainly from European countries. One of my favourites was a project undertaken in Finland by Walter Hackman and Martin Meinander who, over a four-year period back in the 1970s, reared approximately 120 species of fly using 3,700 fruiting bodies (the sporophores) from 184 different species of fungi in pots – that's a lot of pots. And what was the point? Well this research has helped us understand which species consumes what. Why is this important? Well if they have a preference for consuming the ones we like then there could be an economic and/or food security risk.

Of the flies that have mycophagous habits, the majority belong to the Sciaroidea superfamily, with the fungus gnat families comprising all but one family (the gall midges Cecidomyiidae). These families include the 'true' fungus gnats called Mycetophilidae, dark-winged fungus gnats in Sciaridae, Diadocidiidae, Ditomyiidae, Keroplatidae and Bolitophilidae. The last four families are poorly studied and have no common names but we know more about them than a new recently-described gnat family, the Rangomaramidae, which was only described in 2002, originally with only one genus. This has since expanded to comprise 13 genera and 32 species – that's a lot of genera for such a small number of species. It is not the smallest gnat family (Diadocidiidae only has 24 species) but we don't know anything about their biology. They have been collected, killed and identified but our knowledge is limited to just how they look.

This is sadly the case with many species of fly though we can make the assumption that their larvae have similar dietary preferences to the rest of the gnats as they are found in very similar habitats. However, even the fungus gnat families include some non-fungi feeding species (the predacious Keraplatid larvae for example).

Another small gnat family, just 35 species, is the long-beaked fungus gnat in the Lygistorrhinidae family. Once more I can't tell you anything about their biology as most of them have just been identified from bulk samples collected in the field so we don't even have host records for them. They have names, they have descriptions and we are able to work out their phylogenetic relationship to other species, but we can't tell you what they eat, where they live or anything about their behaviour.

Saying that, most people have encountered fungus gnats, even if they didn't realize it at the time, as many live alongside us in our houseplants or greenhouses. I was finally of direct use to my family, thanks to these species, when my sister was looking at flats to buy. As we walked around one prospective place, I spotted some fungus gnats and told her that the flat was more than likely to have a damp problem as these creatures favour dank environments. The survey proved me right. They can also be a right pain if you happen to own a mushroom farm or a plant nursery as they munch away at your produce. Out of all the gnats, it's the dark-winged fungus gnats that are the most economically problematic. They are very easy to recognize, often being black and sleek in form, though, if you ask me, all gnats are quite lovely-looking flies, with their long legs and elegant bodies, and the dark-winged fungus gnats have one of my favourite wing patterns. I find the looping created by the joining of the veins named M1 and M2 rather sensual and become quite syrupy when I look at them.

Over 2,200 species of dark-winged fungus gnats have been described but this figure will probably prove to be woefully small in terms of the true number as there are thought to be many, many undescribed species. As with gall midges, they are very difficult to identify at a species level due to their small size and apparent

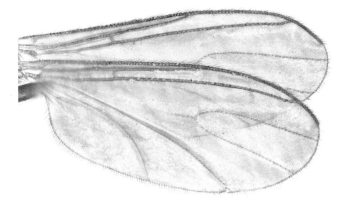

The wings of the dark-winged fungus gnat have a lovely looping M1-M2 vein – one of my favourite wing patterns.

morphological similarity across the family and, as such, they are often ignored in samples! It is estimated that there may be from 8,000 to 20,000 more species out there. Pekka Vilkamaa, a Danish expert on this group of flies, gives the higher figure, citing the enormous numbers still undescribed from his own country, let alone the mega-rich regions such as the tropics. In the UK this family already sits in joint tenth position as the most species-rich group of fly, with 156 described species, though we acknowledge this figure is bound to increase.

Dark-winged fungus gnats are quite cosmopolitan in their appetites with only a few species being really dedicated fungal feeders such as the genus *Lycoriella,* which are global pests for mushroom farms. Annoyingly for farmers, these flies are able to breed all year round and, if left unchecked, multiply to the extent that it would expensive to bring them under control with insecticides. Each time the fly reproduces, naturally occurring mutations in the DNA are passed onto the offspring. Not all of these mutations are beneficial, some may be harmful and may result in the next generation dying out, but some mutations may be beneficial such as being able to genetically resist the insecticides being used. The ones with good mutations are more likely to survive and replicate, so establishing this mutation in the population. Any resistant genes can quickly become fixed in the population, thanks to the fast reproductive rates, meaning producers have to keep searching for new pesticides as the old ones cease to be effective. We are also studying their mating behaviour to disrupt them breeding. One pest species is *Bradysia paupera*, where females waft pheremones to attract males. Once mated no more pheremone is produced and if a male attempts to mount her she kicks the male off with her back legs. We may be able to develop a synthetic version of the pheremone to use in this instance. The larvae of *Lycoriella* tunnel into the stalks of mushrooms and can cause extensive physical and so economic damage. The adults don't help the situation much either as they unwittingly act as transporters to various disease-carrying mites from mushroom to mushroom – some adult flies have been observed carrying up to 85 mites on their bodies! And yes they are still able to fly with this load although I presume not with the same agility or precision as their unhindered colleagues.

Incredibly two species of *Lycoriella* have managed to establish populations in Antarctica at two different research stations. One species made a name for itself within the alcohol store and the other in the sewage facilities. I am more amused than concerned about us introducing invasive species into this new habitat as it's thought they are highly unlikely to be able to survive outside these human-made facilities.

The larvae of another dark-winged fungus gnat species in the genus *Sciara*, *Sciara militaris*, don't stay put during their development but have been observed undergoing mass mobilization. They can form dense processions of thousands of individuals that may reach 10 m (33 ft) in length. We don't know why they do this, but it is

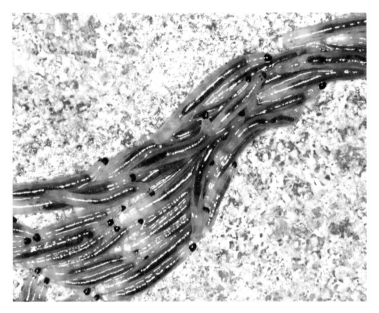

The slow moving, mass-migratory processions of the larvae of the dark-winged fungus gnat, *Sciara* sp. – a poorly understood behaviour – which can reach up to 10 m (33 ft) long.

a phenomenon that has been recorded across the world from the Philippines to Alaska.

Weird locomotory habits are not limited to these gnats but are also observed in the true fungus gnats of the family Mycetophilidae. The larvae of these species nearly all inhabit terrestrial damp environments, where their host fungi are found. The larvae of one of them, *Mycetophila cingulum*, are found in the fruiting bodies of a bracket fungus, *Polyporous squamosus*, commonly called the dryad's saddle. These larvae disperse in a similar fashion to the cheese skippers in that they ping or leap to disperse from the fungus. The pre-pupal stage has been observed leaping from bracket fungi up to distances of 15 cm (6 in). This may not seem like a huge distance but the larvae are only 8 mm (⅓ in) long and so this would be the equivalent of me leaping 3 km (1¾ miles).

Nearly all true fungus gnat adults differ in form from other gnats in being incredibly robust – they all have massive coxa (the top of their legs) and a huge scutum (top of thorax), both packed full of muscles, giving them a very hunched appearance, often with their heads tucked in front of them. Imagine if you can the Incredible

A fungus gnat from the genus *Azana* with their enormous coxa and humped thorax packed with muscles.

Hulk with wings (and maybe not green). To date more than 4,100 species globally have been described and probably that number again are out there waiting to be discovered. This is another problematic group when it comes to applying names to species as many of the females are so morphologically similar to each other that we are unable to figure out which species they belong to. Even the UK's resident expert on this group, Peter Chandler, has issues with some of the females, and many of us mere mortal dipterists will mutter with annoyance about having collected this sex by mistake.

Most species of the true fungus gnats Mycetophilidae are found in toadstools and other fleshy mushrooms, but some are associated with the tougher bracket fungi such as the fly *Sciophila pomacea*. Its larvae live in bracket fungus that grows on plum and cherry trees in the UK. This species had been recorded under the name *Sciophila ochracea* for more than 150 years, a name already used for a different species, but mistakenly used for this one. It was renamed by Peter Chandler in 2006. When we are asked why we collect specimens it is to ensure that we have enough examples

Other genera in this family such as *Phronia* have species that graze on the slime moulds found on decomposing branches next to woodland streams – quite a habitat to live in. The larvae in this genus are mostly free-living and so, for protection, they cover themselves with slime. Three species, though, have utilized another bodily secretion: *Phronia annulata* and *Phronia biarcuata* cover themselves in a thick black slime and *Phronia stenua* creates a hardened black conical covering, known as a test. Both the slime and the test are actually made from faeces! Indeed Carl Steenberg, a Danish snail expert who seems to have become slightly side-tracked, described in a paper in 1924 how he picked tests off the larvae only to see the larvae immediately form new ones by extending their posterior spiracles back and forth over their bodies and laying excretia in a series of waves.

Moving away from the gnats, the crane flies are another group with many mycophagous species. The subfamily Ulinae, in the Pediciidae family, are solely fungus feeders. The whole family are referred to as the hairy-eyed crane flies as, between each unit (called an ommatidium) of their compound eyes, there are indeed short, erect hairs called macrotrichia. These hairs are mechanical receptors but we are not certain why some species have them and others do not. The Ulinae only has one genus, *Ula,* and these differ from all the other species of hairy-eyed crane flies in that their larvae are fungivorous on terrestrial species (all the others are aquatic or semi-aquatic predators). In the UK only two species had been recorded for a long time but this changed in the summer of 2002. The annual summer collecting trip ran by the Dipterists Forum (the UK Fly Society, dedicated to everything 'fly') was held that year in Inverness-shire and dipterists ran around the countryside, sweeping with wild abandon, trying to collect and record as many species of fly as possible. Ken Merrifield and his wife Rita were two such people and they had gone off collecting in Cawdor Woods. Neither is sure who caught the *Ula* specimens but they were subsequently identified by UK expert Alan Stubbs as *Ula mixta*, a species that had only been described 20 years previously and from Slovakia. This species has been happily migrating through Europe and up through Scandinavia and had obviously decided it was time to arrive on British shores. Since those initial records, it has been recorded further south, when it was reared from bracket fungus in Nottinghamshire. It is now seen as an established species and will be added to the British checklist.

From the time the first UK checklist of diptera was published by George Verrall in 1888, we have been continuously adding species to it. His publication, which he acknowledged at the time was a huge underestimation of the true number, listed only 2,881 species. This has subsequently risen to 7,094 and is still far from complete. Both amateur and professional dipterists have been running around the

We have no idea why flies in the gnat family Perissommatidae have four eyeballs separated by cuticle.

countryside for years trying to describe what they find, looking out for invasive species and also studying their biology. Even a country that is thought of as species poor, such as the UK, is rich enough that we still have many unanswered questions.

The Perissommatidae family of gnats could never be described as rich in species as there are are only five, four of these are found in Australia and one in Chile but they could be described as rich in their unusual taxonomic features. The adults are winter fliers which may be a reason why we haven't collected many species (this is the time that dipterists hibernate to their laboratories/bedrooms and spend their time pinning and identifying their summer's hoard). A rather eminent Australian dipterist Donald (Don) Colless, found and described two species in 1962 and then three more in 1969, and placed them in this new family, citing that the genera 'possesses a number of very unusual taxonomic features which, in my opinion, prevent its inclusion in any known family of diptera unless current concepts are drastically modified'. He managed to rear one of these species, *Perissomma fuscum,* from *Boletus* fungi, so we know at least this one is a fungivore. These species are so different to the rest of the gnats that they might even be moved into their own infraorder, Perissommatomorpha. These flies have a feature that is quite unique amongst all flies: they have four eyes! Diptera often have eyes where the facets vary with size across the eye, as they do in the hover flies and horse flies, but these Perissommatidae species have eyes separated by a cuticle and so form four truly separate eyes. We have no idea why!

The Lauxaniid family (Lauxaniidae) may not be a large family, having only 1,900 species, but there have been many morphological tweaks throughout the family which have led to it being very genus-rich with over 200 genera - the differences between the species are too great to be just species changes so they have been described in new genera. This family includes flies that are both beetle and bug mimics and ones that are quite outrageous in appearance. One such genera is called *Cephaloconus*, which translates as cone heads, and they do indeed have remarkable cone-shaped heads.

It is not the larvae that are interesting in this group but the adults as many of them are specialist fungal grazers on fungi found on the surfaces of leaves. The most noticeable feature is the extraordinary morphological plasticity in the shape of the adults, including some that have developed enlarged scraping mouthparts. Stephen Gairmari and Vera Silva, two dipterists from the Americas, published a paper on the rather striking Eurychoromyiinae subfamily of lauxanids which, up to their publication in 2010, contained only one species, *Eurychoromyia mallea*. And what a very odd-looking fly, with what appeared to be corrugated ridges along the front of its flattened face. The word *mallea* means hammer, which gives you an indication of what it looks like. Until this publication not only was there just one species in the group but only four specimens represented that species and these had all been caught in 1913 in the foothills of the Central Andes in Bolivia. None had been found since. Then along came Gaimari and Silva who completely revised this subfamily in 2010 and suddenly five new genera were added (and a further moved from another subfamily). These were collected

The rather bug-like looking *Cephaloconus tenebrosus* – the genus name translates appropriately enough as 'cone-heads'.

by canopy fogging, a technique that involves blasting insecticide into the trees and collecting the rain of insects that tumble down. This may have been the reason why we din't find them as they were hiding high above our standard sampling reach. With the new specimens Gaimari and Silva were able to undertake gut analysis and determine that the flies in Eurychoromyiinae were fungivores as they found fungal spores. And to assist them in grazing, the species have a specialized labellum, the spongy mouthpart, with prongs located along the edges of its pseudotracheal canals which act as raking or rasping structures.

The amazing corrugated ridges on the labellum of *Roryeuchomyia tigrina* which they use to scrape fungi for food.

Perhaps the most deserving of the sobriquet 'fungus fly' are the flat-footed flies of the Platypezidae family which include more than 250 solely fungivorous species. All the species are found in damp woodlands and their larvae are either very host specific or like more variety in their diet. Some species, such as *Agathomyia wankowiczii*, prefer the polyporous (bracket) fungi, whilst others are more attracted to the smellier species found in the Phallaceae, the stink horn mushrooms. The genus *Melanderomyia* is restricted solely to these most pungent of species and, although fungivorous, they may also act as pollinators. Anna Botsford Comstock, founder and original head of the Natural Study Department at Cornell University, wrote about the behaviour of these flies over 100 years ago. She described how the chambers where the spores are 'borne' eventually dissolved into a thick liquid and produced the fetid odour attracting the flies to them. The flies feed and oviposite, and whilst doing so, trap spores on their bushy feet ready to be transported to the next fungus.

We come into contact with the genus *Linderomyia* frequently as the larvae attack terrestrial fungi and are often found alive in soft-bodied mushrooms such as the popular and edible *Agaricus* mushrooms. These mushrooms are not restricted to woodland and so this fly has been caught in meadows as well. The females often aggregate on fruiting bodies to deposit their eggs, and the males can be very striking in appearance as all males in this family have holoptic eyes (ones that join at the top of the head) and generally hang around in swarms.

There is one genus where the swarms exhibit pyrophilous behaviour – they are attracted to smoke. Commonly called smoke flies, all species of the genus *Microsania* swarm and the size of the swarm is determined by the size of the fire. Species have been sampled from wet tropical forests and Peter Chandler, who not only loves fungus gnats but also works with this family, noted that some have also been observed swarming above the tops of trees.

A male from the genus *Linderomyia* showing its holoptic eyes – eyes joining at the top of its head.

We are unsure what it is specifically about the smoke that attracts these species but suspect it is the smoke's chemical composition, although this wouldn't explain the swarming at the tops of trees. A species of dance fly, *Hormopeza* sp., is also attracted to smoke and these predate on *Microsania*, often both males and females flying around in the same swarms as their prey – like shepherds around their flocks.

Some flies have more discerning palates and as such can be very useful to humans. The true truffler or truffle hunter spends years learning their craft. One of the clues that a truffle is nearby is what appears to be burnt vegetation, which is caused by the truffle releasing volatile organic chemicals (VOCs) that kill the vegetation above them. Other trufflers rack or hoe the surface, a method used in the USA and China where the truffles are predominantly found just below the surface. Many people, especially the European trufflers, use sniffer animals such as truffle hogs (pigs) and truffle hounds (dogs). But ever heard of the truffle fly, *Suillia pallida*? Trufflers can often be seen lying down in the woods trying to spot these little creatures and their close relatives. This is referred to as the Mouche method, where the flies are attracted to mature truffles. Both the female and male flies sniff out the truffles and the males attempt to guard suitable sites for females to lay their eggs in. Their swarms may appear as vertical plumes, a jackpot to the seasoned truffle hunter.

I can't finish this chapter without mentioning the scuttle flies or phorids (Phoridae). Yes, once more this ecologically diverse family has struck out in another direction. We met them as pollinators earlier in the book, but they are no stranger to the fungi world either. Of the many species of flies that feed on fruiting bodies, these are a dominant part of the fungi fauna. They formed 7% (third largest) of flies collected from a fungal survey in Switzerland, and they were nearly 7% again (fifth largest) from a similar study in Hungary – their densities varied depending upon the type of fungus. Sadly for me (a massive fungivore), *Megaselia halterata* from this family is a commonly found polyphagous feeder and one of the most important pest species of cultivated mushrooms globally. Luckily money and effort is being spent on researching ways to remove its hungry larvae. It has been nicknamed the tramp species as it has been reported everywhere – on planes, migrating in bird's plumages as well as breeding in nearly anything that has to do with decay. Not only does this species like fruiting bodies, it also likes the decaying slime of fungi associated with decomposing material. Interestingly some *Megaselia* species are parasitoids of the other great mushroom pests, the dark-winged fungus gnat. As of yet we know of no phorid parasitizing a phorid – but with these crazy creatures who knows!

Their fondness for fungi has led phorids down a very odd evolutionary pathway. Erich Wasmann, an Austrian termite and ant specialist, studied a group of flies associated with termite nests. He describes beautifully what he discovered: 'These little termitophile Diptera are indeed a store-house of anomalies whether we consider them from the point of view of a morphologist, anatomist, evolutionist, or biologist.' Steen Dupont, who sadly is now a lepidopterist, alongside one of Denmark's finest fly experts, Thomas Pape, published a review on phorids back in 2009 and identified from literature that there were more than 190 species associated with termites. Some of the species were predators or parasitoids but some had adapted to feeding on fungi gardens or spores that are farmed by the termites. The mating strategies and behaviours of these species are very elaborate and we've already come across similar examples with the decomposer species *Puliciphora borinquenensis,* where males carry ovipositing females directly to the decomposing material that the larvae feed on. The flies that live with the termites cause a massive taxonomic headache and that's mainly due to the females, many of which are wingless. In fact, when they were first described, these females were placed in a different subfamily altogether. They were also thought to be protandrous hermaphrodites – they started off as males and then developed lady bits along the way! No males were found that looked similar to the females so obviously that was the conclusion to draw. There are still many species only described from one sex but, as with *Puliciphora borinquenensis,* once pairs of flies were observed in copulatory flight and the females seen being dropped onto the mounds, these ideas were corrected.

The subfamilies Thaumatoxeninae and Termitoxeniinae contain the species most associated with termites and these flies have undergone some extreme adaptations to live in these environments. Females of the species *Termitophilomyia zimbraunsia*, collected from Zimbabwean termite mounds, rip off their wings once the male has dropped them near the 'cooling' tunnels at the top of the mounds. The female fends off the attack of the workers and smothers herself in the scent of the termite's nest to enable her to penetrate into the main body of the mound where she lays her eggs in the fungus garden on which the larvae feed.

In many species the first instar stage of the larvae is still retained in the egg and the next free-living stages may be very short – some will pupate within a matter of minutes whilst others develop more slowly whilst grazing on the fungi. After emerging from the pupal stage the female remains in what is called a teneral state, where she has not fully matured, for example during the hardening of her exoskeleton. She even mates in this condition and only completes maturation when she crawls into the fungus garden. Often the females emerge from the pupae with their anuses pointing upwards, but these rotate over time to face downwards or forwards. Many species now resemble completely the termites that they live with, having globulous abdomens – a condition called physogastric morphology. Some of these species have pits on their abdomen from which a termite placatory juice seeps. These species are amazingly altered in their morphology but the prize in terms of adaptation must surely go to *Thaumatoxena*, where the female's morphology is referred to as limuloidy after the limulus horseshoe crab which has a very rounded head.

Overall the fungal feeders are one of the least studied groups of what may turn out to be mega species-rich families. Just in the Sciaroidea, the fly super-family to which most of these fungi feeders belong, there are many tens of thousands of species waiting to be described. With luck molecular analyses will help us to distinguish these rather specialized feeders, not just to resolve taxonomic problems, but also to help us understand more about their behaviour and diets.

CHAPTER 8

The predators

'How does Brundlefly eat? Well, he found out the hard and
painful way that he eats very much the way a fly eats. His teeth
are now useless, because although he can chew up solid food,
he can't digest it. Solid food hurts. So like a fly, Brundlefly breaks
down solids with a corrosive enzyme, playfully called "vomit
drop". He regurgitates on his food, it liquefies, and then he sucks
it back up. Ready for a demonstration, kids? Here goes...'

The Fly, 1986

THE ANCIENT EGYPTIANS were a religious and superstitious
bunch, admiring and worshipping many different animals from
the world around them and bestowing different species with different
attributes. We are all familiar with the idols such as cats and jackals,
and maybe even dung beetles – but did you know that they also
worshipped flies? Egyptian soldiers were rewarded with large golden
flies when they showed tenacity and courage, two characteristics of
flies admired by Egyptians. I, too, admire these traits in the many flies
that have predacious lifestyles. I have spent hours lying in wait for
them, watching them attacking often larger and more aggressive prey
species – many insects, snails, crustaceans and, surprisingly, frog's

One of the bulbous eyes of a *Holcocephala* sp. showing its distinctive ridge.
These robber flies predate on the smallest of midges and springtails and so
need excellent vision.

eggs – exhibiting amazing acrobatic skills in the process. Some species even are cannibals.

There are at least 42 families of flies that have predacious species, mostly, but not all, at the larval stage. For example in the sub-family Tanypodinae from the non-biting midges Chironomidae, the larvae all bite. Picture the worms from the 1990 film *Tremors* and you have an idea of what the larvae look like, with large jaws ready to feast on their prey. Predatory flies are found in all environments, from the cold Arctic tundra to sunny tropical beaches. *Beckeriella niger* is a species from the family of shore flies, Ephydridae – its larvae live in the foam nests of Laptodactyline frogs. It has been observed consuming not only the eggs, but also the tadpoles.

In the Chaoboridae family, commonly called the phantom midges, the larvae of *Chaoborus edulis*, referred to as glassworms, are predators of plankton and are found in enormous numbers at the bottom of lakes. Amazingly they capture their prey with their antennae which have been modified to become prehensile! On top of that, as they live at very deep depths they have reduced their breathing system to just two air sacs. These further act as buoyancy aids and can shrink or swell on demand. These amazing creatures no longer breathe aerobically but instead use an anaerobic cycle to generate adenosine triphosphate (ATP) – essential for energy production. These adaptations ensure large numbers can survive and chaoborids form massive swarms as adults, millions upon millions emerging at once.

Oedoparena glauca, from the family Dryomyzidae, a common coastal fly from central California all the way up to Alaska, is a predator of barnacles. Both the adults and the larvae live in the intertidal region, but once more it is the larvae that are the specialist predators. The adults assist by depositing eggs on the inner surface of the barnacle – the operculum – that seals the entrance to the barnacle when the tide is low and the barnacle exposed. Once

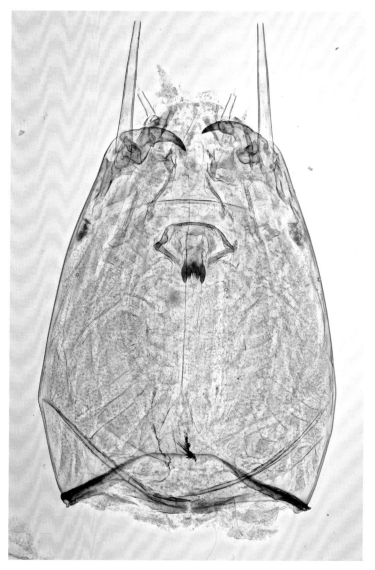

The head capsule showing the teeth of the predacious non-biting midge,
Heterotrissocladius grimshawi.

hatched, and when the barnacle is open, the larvae enter the main chamber. Barnacles spend nearly three quarters of the day exposed and therefore closed, and it is during this time that the larvae munch away in safety. Larvae can move and consume two to three barnacles before they pupate, again in the safety of the barnacle's shell.

Once again back to the scuttle flies. One species, *Megaselia mallochi*, has larvae that predate on the dark-winged fungus gnat, *Bradysia confinis*. This scuttle fly has also been shown to eat the gnat larvae even if it has been parasitized by the ichneumonid wasp, *Stenomacrus laricis* – a double protein packet and, just think, one more parasitic species and you have the fly equivalent of a turducken. (Note a turducken is a culinary delight apparently of a deboned chicken, stuffed in a deboned duck, stuffed in a deboned turkey.)

Other species don't bother to hunt for their prey, instead they let their prey come to them. The small family of flies Vermileonidae, or wormlions, are closely related to the bigger snipe fly family Rhagionidae but their larvae are completely different in terms of how they look and feed, so justify being a separate family. Wormlions are named after antlions, a completely different type of insect but one which has a very similar method of capturing prey. Picture the scene in *Star Wars: Return of the Jedi* featuring the Great Pit of Carkoon. The large, inverted conical pit in the film was home to an adult sarlacc – 'a large, omnivorous arthropod with long tentacles that rimmed its mouth'. Change this description from arthropod to larva and remove the crazy spines and tentacles and you have a wormlion larva. They are solitary individuals with only one larva per pit. Any poor creature that stumbles into the pit meets its doom as the larva flicks sand onto the walls causing them to become unstable and then collapse, hampering any escape attempt and forcing the prey towards the larva lying across the bottom. The larva wraps itself around its victim, holding onto it with a pseudopod, a fake leg. It then sinks fan-like mandibles into it. It's thought these larvae are also venomous and Justin Schmit, an American entomologist who works on venomous insects, notes that the docile behaviour of the prey once attacked by the larvae is similar to that of species attacked by robber flies, a known venomous group that subdues its prey. The wormlions are not always alone in their pits. The larva of *Scaptio muscula*, a horse fly, sometimes lies in with the antlion and joins in the feasting.

Predatory mosquito larvae like a mixed bunch of habitats, with some developing in permanent water bodies like ponds, some in temporary sites like tree holes, and some in rather more unusual situations, such as inside pitcher plants. Normally thought of as a very hostile environment, there are several species of mosquito larvae that can develop in the water body of the pitcher, the phytotelma, including the pitcher plant mosquito, *Wyeomyia smithii*. The fact that mosquito larvae can survive, living and feeding on insects that have fallen into pitcher plants is one thing, but why doesn't the plant eliminate them as surely they are eating the plant's food sources? The answer is that pitcher plants need them. Initially the plants produce a large amount of digestive enzymes to enable them to break down dead prey. But as they age, they produce fewer and fewer of these enzymes and so are no longer able to feed on the larger items of prey. The mosquito larvae are voracious predators, devouring everything that falls into the pitcher, but they are also rather messy, shredding their prey and scattering entomological crumbs around. This debris is small enough for the ageing plant's meagre supply of enzymes to work on so ensuring that the plants survive – a very amenable collaboration between two predators.

There is a strikingly good-looking group of flies called marsh flies (Sciomyzidae), which have long faces and thick long antennae protruding out of the front of their heads. They are also sometimes called snail-killing flies as some do exactly that. Within this family there are parasites and parasitoids but also a healthy number of predators. Some species have larvae that are very good at hunting and devouring snails and other molluscs such as slugs and pea clams. Interestingly it is the presence of fresh faeces or mucus that attracts the adult flies. Some also prey on earthworms. Most of the prey though are found in aquatic or semi-aquatic habitats, and to cope most sciomyzid larvae have a large abdominal disc which they hold above the water's surface and through which they breathe – maggots breathing out of their anal appendages is a very effective but standard trick.

Even with the disc held out of the water, the larvae are still flexible enough to carry on hunting and feasting under the surface. We are beginning to value these flies in the field of biological control as they can kill snails which are hosts to terrible parasitic infections such as schistosomiasis. Also known as bilharzia, this is a disease caused by schistosoma parasitic worms. If left untreated it can result in kidney failure, liver damage and infertility, and has a huge impact on people who contract it. In Africa infection rates are second only to those of malaria. Sciomyzid larvae prey on the intermediate hosts of this parasite, snails from various genera but particularly *Biophalaria*. This prevents the parasite from continuing its life cycle in us. This is a much cheaper and more permanent way of preventing people getting infected with these parasites than either chemical spraying or developing drugs for prevention or cure.

Flies also came to the rescue of a parasitic problem back in the 1960s in Hawaii. *Fasciola gigantica*, the liver fluke, was a parasite infecting large numbers of dairy and beef cattle, and in some areas the infection rate was as high as 87%. This had a considerable impact on the welfare of the cattle and on the industries that depended on them. The intermediate host of this fluke was the freshwater snail *Lymnaea ollula*, which lived everywhere. The US Department of Agriculture and Conservation in Honolulu decided to intervene and solicited the help of Clifford Berg and Stuart Neff – entomologists who had a good track record of working with snail-killing flies. Research led them to first recommend introducing the Central American fly *Sepedomerus macropus*, commonly called the liver fluke snail predator fly (maybe one of the longest common names but very apt). This species was great at preying on the snails when the snails were submersed, but sometimes these snails undergo a terrestrial stage, when ponds dried up for instance. So another fly was needed to target this stage. Berg and colleagues looked at some European species, and found one from Denmark called *Pherbellia* (*Chetocera*) *dorsata*. Five individuals were flown to Hawaii – yes, this sort of thing actually happens. Although they all got on well with each other and mated, no offspring were successfully reared. So more were sent for and 80 puparia were soon winging their way across the Atlantic. Of these, 61 adults were reared and these went on to reproduce. After three generations they had more than 12,000 flies that were introduced into the environment. Not only were the flies successful in killing the intermediate host but studies of these newly released flies in the wild showed that they had very little impact on the native endangered snail populations or the native carnivorous species preying on bad snails.

Snail-killing flies are not only useful for eating snails. Their potential is also being investigated for the biological control of pestiferous slugs such as the grey garden slug. Trials are being done to determine whether species can be released to control other agriculturally devastating pest species. One such species, *Tetanocera elata,* is being closely studied as an important enemy of slugs. Along with many other fly species, sciomyzids are venomous and probably one of the least researched of venomous species.

People don't generally associate flies with being venomous, but in fact many families of flies have venomous members. Brian Fry, a venom specialist, and colleagues published a substantial paper in 2009 on animals that produce venoms. They found that the protein composition of the saliva of both venomous flies and bloodsucking ones was so similar that they regarded the latter species as producing saliva that was a special subtype of venom. So when we are being attacked by mosquitoes we are receiving tiny venomous bites.

Adverse conditions can encourage the evolution of predatory flies, as generally hardier animals can endure more just to stay alive. Caves can be pretty adverse environments – they are dimly lit, damp and there is very little food irrespective of what you feed on. The Keroplatidae family is mainly forest dwelling, but many of the species in this family that live in caves have evolved predatory habits, and several genera have very particular adaptations for catching prey. *Arachnocampa*, a small genus of eight species found only in New Zealand and Australia, has very predatory larvae. Like their cousins, the true fungus gnats, the larvae produce silk strands which they dangle from their central nest of silk on the ceiling of the cave, but unlike their cousins these act as snares. Up to 70 strands hang down from one nest and these can be very long – up to 40 cm (16 in) – and very sticky with beads of mucus adorning their lengths.

Now, as I am sure you realize even if you haven't been in a cave, these environments are dark, very dark. To make themselves, and their strands, more alluring to their prey, the larvae bioluminesce and glow with different intensities depending on how hungry they are. Any flying insects that are attracted by the twinkling 'stars' get caught on the silky snares. The gnats then pull up their sticky victims and devour them. As already mentioned in Chapter 1, *Orfelia*, another Keroplatid genus, has further improved this fiendish death trap by adding poisons to the strands. The snares include ampules of oxalic acid that rupture when touched and start breaking down the trapped prey before the larva has even touched it.

It's not just larvae that can be venomous. There are some seriously impressive adult predators in terms of behaviour and physiology, especially in the robber fly (Asilidae) family. Work is only just beginning to properly study the venom glands of these flies – found both in the adults and the larvae. The collection at the Natural History Museum contains not only many thousands of specimens of robber flies but also the prey on which they were feasting when caught. One of my favourite drawers of robber flies has, nestled in-between a male and female specimen of *Mallophora infernalis*, a long-horned grasshopper and a tiny note stating: 'These two were caught in flight carrying the above orthopteroid (still alive)'.

Who knows what was going on with these two – was the male trying to attract the female with a nuptial gift? Or had the female caught the bush cricket and the male got lucky with both a meal

The beautiful but deadly strands of mucus used by *Arachnocampa* sp. of fungus gnats to catch their prey.

These two flies were caught in flight carrying the above Orthopteran (still alive).

These male and female specimens of the robber fly, *Mallophora infernalis*, were caught in flight carrying their prey, a bush cricket.

and a lady? Turns out it was no-one's lucky day except for the entomologist who caught them! The collection contains many different prey species, including beetles, other robber flies and dragonflies – supposedly the best insect fliers, but obviously not so to these little predators. Some robber flies can be very robust, while others are slimlined – but all of them have a moustache (even the females) called a mystax. The mystax plays the essential role of protecting the fly's delicate mouthparts from the flailing limbs of its prey. *Mallophora* robber flies have quite a reputation for being among the more hardcore predators in this family. This genus is sometimes referred to as the bee killers as many have been observed preying on bees, wasps and their relatives. Honey bees in particular are easy prey as they are fairly slow fliers, a good size for the flies and locally abundant – hives can contain tens of thousands of bees. There have been reports from Florida, USA, of these robber flies having an actual economic impact due to the large numbers of bees they kill. *Mallophora orcina*, the southern bee killer, has been reported attacking hives in their hundreds.

There are even tales of these flies killing hummingbirds. Ed Johnson's post on the Texbirds ListServ from 2008 details how, for over half an hour, he watched a fly feeding on a ruby-throated hummingbird – the bird was grabbed, subdued and then had its insides sucked out. The species was *Mallophora leschenaultia*, which

The most robust and fluffy robber fly, *Mallophora leschenaultia* – a formidable predator that has been recorded attacking prey such as hummingbirds.

has two common names – the black bee-eater or the more sinister Beelzebub bee-eater – and they are more frequently seen predating on bumblebees, honeybees and wasps. Although smaller than hummingbirds, this one obviously has no qualms about attacking prey larger than itself. And this is not the only species of robber fly that has been seen attacking hummingbirds. There are several accounts of attacks, including one by the macabrely named red-footed cannibal fly or bee panther, *Promachus rufipes*.

Robber flies are one of my favourite groups of flies and when asked what my favourite species of fly is – which to be fair does change depending on what I've discovered about the species I'm currently working with – the species I name mostly is the bumblebee robber fly, *Laphria flava*. This species is so majestic I have taken many an opportunity to watch it on field trips in Scotland.

Bumblebee robber flies are found across Europe and Asia and, like all adults of this family, are solely predacious. Their mouthparts, as with all robber flies, have a hardened tube-like proboscis that surrounds a very long and slender hypopharynx, which is a tongue-like structure. This forms the lance-like tube that penetrates the prey. They feed on beetles, thrusting their mouthparts through a weak spot – attractively this is often the eye – and releasing saliva containing digestive enzymes and venoms into the poor beast. A fly can catch something on the wing, subdue it, suck it dry and then discard the remaining husk in a relatively short time. Efficient feeders! Bumblebee robber flies are, like all *Laphria*, and as their common name indicates, bumblebee mimics, but others across the family mimic different kinds of bees, social wasps and parasitic wasps.

Species within the genus *Holcocephala,* a name that means grooved head, do indeed have the most pronounced grooved head in the robber fly family. They also have massive pronounced eyes that are not just for show. They specialize in predating on midges and springtails – some of the smallest creatures around – and so they

The bumblebee robber fly, *Laphria flava*, quite possibly my favourite fly, with a hardened proboscis for penetrating prey, often through the eye.

need excellent vision. All robber flies have very advanced vision and their eyes are constructed out of thousands of ommatidae, or photoreceptive units – the facets on the eyeball. The ommatidia can be different sizes and they enable the flies to have a large angle of view and detect rapid movement. Have you ever wondered why, when you have gone to swat a fly, it so easily escapes? Well they can see the swift movement and it is often the opposite that actually fools them – just move your hands very slowly towards them and you can catch them more easily, obviously then releasing them unharmed.

Another personal favourite is *Wyliea mydas,* a robber fly from the subfamily Asilinae. This species is a wonderful mimic of the

Three predators? No, two predator flies *Wyliea mydas* (bottom left) and *Pepsis formosa* (top), and the actual *Pepsis* sp. wasp (bottom right) they mimic.

genus *Pepsis*, a tarantula hawk wasp, a very large parasitic wasp in the Pompilidae family, the larvae of which develop in tarantulas. This robber fly's similarity to the wasp is remarkable. Although the robber has a small proboscis similar to the wasp, in common with all flies it doesn't have a sting at the other end. The sting is a modified egg-laying tube only found in hymenoptera (and obviously only in females). However the fly still physically mimics the wasp's sting by waving its own genitalia around in a stinging motion. The *Pepsis* must really be something considering robber flies themselves are so venomous and agile you wouldn't think they needed to mimic another predator!

Wyliea mydas is not alone in mimicking the tarantula hawk wasp. There is another species in the Mydas family (Mydidae) that also mimics it. In fact it is very difficult to tell all three species apart. *Mydas xanthopterus* adults look practically identical to both *Wyliea mydas* and the wasp. However, this family doesn't have predaceous adults and isn't venomous so it makes more sense for it to mimic a more dangerous species.

One of the largest flies known to science is *Gauromydas heros*, also one of the Mydas flies, which has a body length of up to 6 cm (2 in). These are very robust flies as adults and have equally impressive larvae that can reach 4 cm (4½ in) long. Don't think just because the adults are vegetarian that the larvae are harmless – they are formidable predators. Adult females are not thought to eat at all but instead rely on fatty substances that are stored in their abdomen, laid down during the larval stage. *Mydas* larvae, as with most dipteran larvae, have been poorly studied and we are still unsure about much of their biology. The literature states that some species deposit eggs in sandy soils and others in decomposing wood, but all of them apparently feed on beetle grubs. The larval feeding habits of *Gauromydas heros* have been studied the most in this family and the larvae are thought to prey on *Coelosis* sp., a genus of horned dung beetles that act as the waste disposal team in leaf cutter ant nests, living in and consuming the ants' waste.

Another highly predacious group of families are the Empidoidea, a superfamily of four or five closely related groups that include the long-legged flies (Dolichopodidae) and the dance flies (Empididae). The Dolichopodidae, or dolis for short, are a very common and widespread family of flies in which all of the adults and most of the larvae are predatory. The adults have a slightly unusual way of feeding compared to the other families in Empidoidea. Instead of thrusting their mouthparts into prey and sucking out the dissolved insides, as the robber flies do, some of them capture and hold their

Gauromydas heros – one of the largest and most impressive of all flies.

prey with their mouthparts and then shred them, which causes the internal contents to ooze out. They have an upper lip or labrum and a lower lip or labium, and the prey is held between these. Just under the lower lip there are a series of surfaces called epipharyngeal blades, which are used to shred the prey. These blades are one of the many features enabling us to identify different genera across the family.

When the males of this family are not hunting or feeding they carry out some of the most amazing courtship displays to woo the females. Some species exhibit extraordinary male-to-male pursuit, vigorously defending their territories before turning their attentions towards the more amorous task of trying to attract females.

To assist in hunting, territory defending and courtship, some species are incredibly fast fliers. *Poecilobothrus nobilitatus*, a fairly large and distinctive doli, not only flies fast but research has shown that it is able to change its direction of flight within 15 milliseconds. And yes 15 milliseconds is incredibly fast. All flies are able to act on what they see much faster than most animals and appear to be capable of responding, what appears to be almost instantaneously to different stimuli. Rather than this information being processed in the head region, it gets passed almost directly to the wings, cutting down on thinking and processing time. This is not only useful in hunting, or escaping from being hunted, but also in mating rituals.

The male of *Poecilobothrus nobilitatus* is also able to quickly change direction to enable him to perform complex courtship rituals on the ground, where he keeps at a constant distance of 2.5 cm (1 in) from the female. A common species found across Europe, it didn't have a common name for many years until in June 2012 *The Guardian* newspaper in the UK ran a 'name the species' competition in which this species was included. The following month Alan Thomas was announced as the winner of this fly's common name and can be proud that the name 'semaphore fly' is now used in fly conversations up and down the country. The name is most apt as the wings of the male have distinctive white tips, and to woo the females he buzzes his wings perpendicular to his body. His wings are proportional to his body size – the larger his thorax the longer his wings.

Across the whole family it is not unusual for the males to have wings with spots as well as more ornate features such as expanded feet or paddles, decorated wings, silver patches and expanded antenna, all of which are used in courtship. They also have some of the largest, most obvious genitalia of all flies. Unlike many flies, which tuck their genitalia away, these ones are always proudly displayed. A favourite of mine is the genus *Dolochopus* where many species have massive genitalia – sometimes the size of their abdomen.

Very complex courtship displays seem to be a characteristic of most predatory flies. A closely related family to dolis are the empids or dance flies of the family Empididae. They almost put the dolis to shame with some very elaborate routines. Unlike the dolis, not all the adults are predators and, of those that are, it's often just the males that hunt. In the subfamily Empidinae few females hunt, instead relying on the males to do the work and allowing themselves to be wooed with prey items called nuptial gifts.

A male *Plagiozopelma* sp. with their incredibly ornate antennae.

Different species have evolved different delivery methods for these gifts. In the basic method, prey items are brought intact to the female by the male. In some species, the males swarm, enabling them to be more visible to the females, while others perch individually on branches, holding on to freshly caught prey for the female to feed on during copulation. Often the females finish this gift and leave before copulation has finished. The males could find larger prey as gifts to prolong the act, but this would take more time and effort. Instead some species take to wrapping the gifts like you would a box of chocolates. *Hilara*, a commonly distributed genus, has huge swollen 'feet' on its foreleg. These swellings contain the glands used for silk production, the wrapping paper for the box of chocolates. Adult males hunt for prey, wrap them up and present the gift to the females. However, if the females take a long time opening and then feeding on their gift, the males often finish their business before she has finished eating. Not wishing to waste excess food on the female, individual males have been observed taking back any half-eaten gifts and rewrapping them for future matings, rather like a date turning up with a half-eaten box of chocolates. In the genus *Empis* many of the species just present empty frothy balloons, made out of saliva and nothing else, not even a mangled bit of prey stuck to the side of the balloon (this too has been observed). Poor deluded female, but hats off to the male.

That's not the worst of it for the females of this family though. Males in the genus *Rhamphyiomyia* don't present courtship gifts at all. They still hunt for the females, but they make the females work for their dinner – or should I say dance. In this genus the females have evolved secondary sexual characteristics – an unusual occurrence in nature as it is usually males that have the exaggerated characteristics used in courtship. Firstly, the females have pinnate scales on their legs, which make their legs look very hairy. They have also developed oversized, bulbous-looking wings. And the final

Hilara sp. with their enlarged swollen leg segments where the silk glands are found.

piece of sexual armoury is the development of eversible abdominal sacs – this means they can quite literally blow up their backsides. Eversible sacs can be inflated and deflated as the female requires. They wrap their hairy legs around their abdomens and pump air in and out of it during flight. This is thought to be a way of showing the males how much room they have in which to develop eggs. Next time you see females posing on a dance floor be grateful that humans generally don't go to these extremes. I don't feel too sorry for the fly ladies though, as the males of some species have a huge twisted phallus that has been described as resembling that of a buzz wire toy.

But we can't let the males have it all their own way. Finally, let's consider once more the biting midges from the family Ceratopogonidae. This family further subdivides and within the subfamily Ceratopogoninae there are bloodfeeders as well as insectivores. Of the insectivorous species, the females generally prey on many species of nematocerous flies while mating. However in some species the females have developed a quite wonderful mating strategy. *Serromyia femorata* is one such species. The female lets the male initiate copulation but then, in the process, she pierces his head, releases digestive enzymes that dissolve the contents and then sucks him dry! Nothing is left of the male but an empty exoskeleton, which then breaks off at his genitalia (remember they are still in the act of copulation) forming a natural but temporary plug to try to prevent her from going off and immediately copulating with another male. I guess if you are going to die this may be one of the nicer ways of going!

So the predators are the naughty ones, the vicious ones. But they are also the helpful ones. Many species prey on insects that we consider pests, including snail-killing flies that also kill slugs and many species of hover fly larvae that specialize in feeding on aphids. Just take some time to sit by a pond watching a species of doli and become entranced with their behaviour – way more entertaining than your average nightclub.

CHAPTER 9

The parasites

Every man has inside himself a parasitic being who is
acting not at all to his advantage.

William S. Burroughs

O H WHERE TO BEGIN with these species, the stuff of nightmares, the parasites, and the even more specialized parasitoids. But to me these are some of the most intriguing and attractive of all the flies. I've encountered many a dipterist who has regaled me with stories of their own, very personal interactions with a parasitic maggot. One had a bot fly 'guest' for about a week before his wife finally made him get rid of it as she could hear it munching away during the quiet hours whilst in bed. I often lament not having one though this thought inspires looks of disgust rather than sympathy from friends.

Parasites are defined as species that, at some stage of their life cycle, live in or on another organism, obtaining nutrients, water and sometimes oxygen from the host species. Many of the species of diptera that I termed vegetarian – the borers, miners and the like – are technically parasites, as the host plant suffers damage, sometimes enough to cause its death. But it's the maggots living in animals that get our imagination going.

How could you not love the *Cuterebra emasculator*? Well, I guess if you are a squirrel you wouldn't as the fly's eggs hatch under the squirrel's skin.

Parasites may or may not cause the death of their host but when they do they are called parasitoids. We find parasitic stages in both the larvae and the adults, but it is only the larval stage that are parasitoids. There are many species of parasitic (and parasitoid) animals, especially within the insects, with wasps generally considered the most impressive. In 1997 Donald Fenner and Brian Brown, two American entomologists, estimated that 78% of all parasitoids were hymenoptera – ants, bees, and wasps – with flies coming second with 20%. (However, if proved correct the previously mentioned gall midge study is going to completely alter that figure.) All wasps that are parasitoid have evolved from one common ancestor but this is not the case with flies. There have been at least 100 different instances in which fly species or species groups have turned parasitic and this feeding behaviour is found in at least 31 families of fly – a term called recurrent evolution, where there is repeated uptake of a particular trait. Although flies lose out in terms of the total number of parasitoids in comparison to wasps, they have the last laugh as many of them have taken to having hymenopteran hosts.

This is not the first time that parasites have featured in this book. I briefly discussed bee flies and their unusual parental care on p.29 – they wrap their eggs mid-air in ballast and hurl them into or close by solitary bees' nests. Bee flies are some of the most amazing of parasitoids and every year, with the onset of spring, the Museum is inundated with reports of tiny fluffy flying narwhals from gardens across the UK. The sightings are actually of the large bee fly, *Bombylius major*, whose hairy body and long proboscis give rise to these inquiries. I have watched these creatures for hours, the harmless adults feeding, trying to find a suitable host nest for their parasitical offspring, or flirting in the morning sunshine.

Another family with a liking for hymenopteran hosts are the thick-headed flies in the family Conopidae which, as their common name suggests, are a very distinctive-looking fly in their adult form, being more angular and sharp edged than most dipterans. These species are often notable mimics of bees and wasps, presumably to enable them to sidle up to their intended victims discreetly. In many species from this family, the female is not fussy about her larval host and will attack a range of bees and wasps for her polyphagous offspring. She catches the host mid-air and forcibly inserts an egg into it. Many of the females have modified their abdominal segments into what is termed a theca, commonly referred to as a can-opener, which is used to prise open the abdominal segments of the host species. Theca vary from species to species and so can be used as one of the features to identify them. It makes a refreshing change in diptera to identify species based on female genitalia rather than male.

The deadly thick-headed fly *Conopid* sp. which looks like, and parasitizes, bees and wasps.

Only one egg is laid at a time in each host. Once the larva has hatched it sets about consuming the insides of its host, ultimately causing the host to die. The larva then takes advantage of this empty body, using it as a shelter during pupation. Using the host to protect pupae is a common and sensible practice employed by many a fly and it saves them from having to use up valuable food reserves in either constructing a tough case or seeking out an alternative hiding place.

Not all thick-headed flies prise open their hosts. Some species have modified the egg into the shape of a harpoon. Within the subfamily Stylogastrinae or Stylogastridae (there are still arguments as to whether this is a subfamily of Conopidae or a family in its own right), there is the genus *Stylogaster*. The larvae of this genus specialize in parasitizing cockroaches, crickets and the larger chunky flies including house flies and tachinids which, ironically, are parasitic flies themselves. *Stylogaster* species are often associated with army ants and have a cunning way of using the ants to find potential hosts. As the colony of up to 15 million army ants maraud around, they flush out all the hidden insects and these are then chased after by the flies. Females either stab or shoot out little egg darts at speed towards the moving target – a process that rarely allows for a great deal of precision. Eggs have been found all over the hosts including in their eyes.

Looking at house flies of the family Muscidae, adults and larvae have a range of feeding strategies including some parasitic species. Birgit Fessl, who works for the Charles Darwin Foundation in the Galapagos Islands, has been working on parasitic infestations on the Darwin's finches and other songbirds. She, alongside other researchers including a Canadian dipterist Bradley Sinclair, is studying the invasive house fly *Philornis downsi*, which is having an enormous impact on the finch populations. This is an odd species of fly. The first instar stages are endoparasitic, more accurately described as agents of myiasis (parasitic infestations caused by flies), and are found in the nostrils as well as under the skin of nestlings. The late second and third instar stages are free living; having exited their hosts they now live in the nest material. These maggots are ectoparasites feeding on the blood of the nestlings at night. Recent models determining the impact of these flies on the finch population

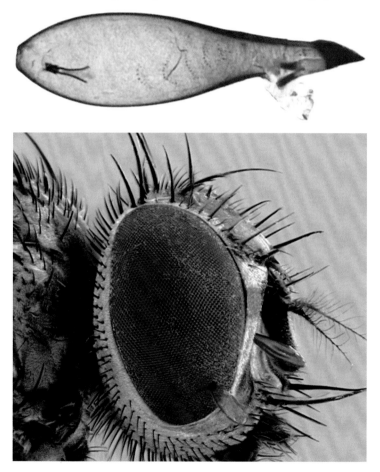

The harpooned eggs of *Stylogaster* sp. pierce right into the eye of its muscoid victim.

since the flies' arrival in the 1960s predict that the birds will go extinct within 50 years if nothing is done. Don't worry though, Fessl, along with many others, is working hard to save the birds from their current plight. It's rather ironic that these finches, which were so influential on Darwin and his theory of evolution, are now not able to evolve fast enough to cope with this introduced parasite.

Tachinid flies belong to the large Tachinidae family – more than 10,000 species have been described so far – and all the species have parasitoid larvae with mostly insect hosts. This family may end up being very useful in pest control as, unlike many conopids, they are very host-specific. I was once involved in a biological control project where I recorded and imaged the larvae of the tachinid *Medina collaris* as they crawled out of the abdomen of their pest species, the heather beetle, *Lochmea suturalis.* On one occasion I watched one crawl back in – obviously not ready to pupate in the outside world. Densities of this beetle can reach very high figures and have a huge impact on the heather that they feed on, eating both young and old plants. In turn, all of the species that depend on the heather are also affected. Of more concern to me is that heather plays a key role in flavouring whisky and we can't let anything impact on that!

The adults in the Tachinidae family are very bristly, indeed they are some of the spikiest adults you will ever come across. Quite why they should be so bristly is unknown but the *Epalpus* genus is so bristly that many species in it are referred to as hedgehog flies. In some species of tachinids the adults have also developed 'ears', more precisely tympanal hearing organs, to enable them to locate hosts for their offspring. These tachinids, as well as some species of flesh fly, are able to listen to the calls of crickets, grasshoppers and bush crickets as well as those of some true bugs such as cicadas, though this is not too difficult with the latter given how noisy they are. The tachinid subfamily Ormiini, a rather small group of 68 species predominantly found in the tropics, especially South and Central

America, are the only flies so far reported to have true tympanal hearing organs. The female of this subfamily has what look like shoulder pads and these expanded shoulders accommodate a rigid frame for her tympanal membranes. These are home to the sensory organs and larger air spaces, called tracheal spaces, which enable her to pick up airborne sounds. The egg-laden female flies around eavesdropping and can hear the mating calls of the grasshoppers and other hosts suitable for her nefarious needs.

Not all prospective mothers are so well equipped with tools for locating the hosts for their offspring. Many lay already incubated eggs that hatch immediately into hunting larvae, while others simply lay their eggs on plants in the hope that these will be consumed along with the leaf by some herbivorous insect.

A very bristly hedgehog fly, *Epalpus* sp. of the Tachinidae. Quite why they are so bristly is unknown, but they are some of the spikiest flies.

Parasitic living may be great in terms of having your food on tap, but there are other issues that need to be taken into consideration such as being able to breathe. To achieve this, some species have learned to break into the tracheal or breathing system of their insect host. All insects have a network of internal pipes called tracheae that enable air to migrate passively around their bodies. The parasites simply lock on to these pipes to ensure a continuous supply of air for themselves. A slightly more sophisticated approach has been adopted by the larvae of the tachinid *Exorista larvarum,* which bury themselves in and feed on the caterpillars of various moths and sawflies. These larvae have simplified mouthparts that they use to macerate and penetrate the walls of the host. This causes an immunological response in the host that makes the protective walls darken and harden through a process of melanization around the entry point. So as the larva penetrates into the body, a thin tunnel is produced in its wake. The posterior, breathing end of the maggot remains in this funnel, held on by an anal hook, to obtain air whilst the rest of it gorges on the doomed host.

Although we are still in the early stages of studying the use of tachinid flies as agents of biological control, other families have already been identified as exceptionally useful in our fight against agricultural pests. One such family is the scale flies Cryptochetidae. All 33 described species have endoparasitic larvae feeding on the insides of the true bugs from the family Coccoidea. This superfamily of bugs contains many serious economic pest families including scale insects and mealy bugs, with the former attacking field crops while the latter prefer citrus crops. These bugs are not big themselves, and the flies that attack them are even smaller – some only reaching 4 mm (⅙ in) in size, but their diminutive size does not detract from their economic value.

A scale insect of major economic impact is the cottony cushion scale, *Icerya purchasi,* which has a very polyphagous diet but has a penchant for citrus plants. Originally native to Australia, this little creature, which resembles a tiny beetle, was accidentally introduced to the USA from New Zealand on the bark of an acacia tree in California in 1868. Released from the controlling constraints of its natural enemies, it went on a feeding rampage and, in less than 20 years from its initial arrival, it was decimating southern California's citrus industry. Huge economic hardship was already affecting the region, and this prompted the US government to step in. Charles V Riley, Chief of the US Department of Agriculture Entomology Division at the time, was the lead investigator determining what this beastie was, where it came from and what on earth they could do about it. Traditional methods, including cyanide, had no effect on this hardy creature, and vast amounts of money were spent ineffectively. Riley suspected that the pest may have come from Australia as there were many agricultural and ornamental imports at that time from there. He hired the entomologist Albert Koebele to investigate, and then began a correspondence with an Australian entomologist, Frazer Crawford. They searched for natural enemies of this pest and, not long after their initial contact, Crawford determined that *Cryptochetum iceryae,* the scale fly, was a predator/parasite of the bug. In field trials, he demonstrated that it had a serious impact on the scale bug populations in the region of Australia surrounding Adelaide. Koebele high-tailed it across to Australia, and took the scale fly – and a ladybird that had also been shown to be effective at control – back home with him. Both of these creatures had a dramatic immediate impact on the pest populations and within months it was reduced to harmless numbers.

Even though they were able to determine that *Cryptochetum iceryae* was a natural enemy of this species, it often takes years to understand the life history of flies. Emil Hans Willi Hennig (1913–1976) was an amazing biologist and was important in the development of evolutionary theory. Hennig is considered to be the father of cladistics, which is a way of describing how things are related to each other based on shared characteristics. But more importantly for this story he was a great dipterist. In 1952, more than 50 years after the naming of the *Cryptochetum iceryae* fly species, he wrote that: 'the larvae of the genus show much that is remarkable and also much that is not solved yet'. At that point we were unsure how many instar stages the larva went through (we now know it is four), and know nothing about their development. He did know, though, that the pupae only blew off one lid when 'busting' out of the puparium, the method of emergence that all Schizophora flies use to exit the puparium, rather than the two lids used by closely-related species. Today we know slightly more about the life cycle of *Cryptochetum iceryae* including which larval stage attacks the host (the second) and whether it attacks larvae or adults or both (both), and this knowledge helps in fighting infestations.

Another more unusual group of parasitic flies is one that has adapted to living on frogs, and they are commonly called frog flies. Only found in Australia, the larvae of a genus *Batrachomyia* from the frit flies family Chloropidae parasitize a number of different frogs, developing in a protected environment under their skin. Because of the loose nature of frog skin you can easily see the lumpy protrusions of larvae underneath it, and there is often more than one larva per frog. The larvae can be enormous in relation to the size of their hosts (and presumably quite irritating because of this). A paper on the genus by Francis Lemckert, an Australian zoologist, describes one larva as 'extending about 70% the length of a frog. On removal, one of the largest larva was found to represent 7.1% of that frog's remaining mass.'

Endoparasitic flies – those living inside their hosts – are not common across all vertebrates and are mostly associated with mammals. It is very unusual for a mammal host to die as a result of these infestations, but obviously there are bound to be some negative effects on their health. The poor frogs in the previous story often died after

The larvae of the camel bot fly, *Cephalopina titillator*, with the lighter earlier instar clearly showing its mouth hooks

the parasites emerged because of blood loss from the open wounds. Some of the most charismatic endoparasites are the bot flies in the family Oestridae. There are four subfamilies in this family, one of which, the Oestrinae, contains the nasal bot flies – or more affectionately named the snot bots. The camel bot fly is one such species and is found in the nasal cavities of these large mammals, predominately in Africa, but they have turned up in camel populations elsewhere in the world. The larvae, which look like orange-coloured rainbow drops (puffed wheat sugar bombs from my childhood), are called *Cephalopina titillator*.

You would have thought that with a species name like that, the camel would gain some pleasure from these little creatures, but as far as I know there is nothing remotely stimulating about these maggots. Instead they probably irritate the camel beyond belief. Upon exiting the female fly, the eggs immediately hatch into maggots and these are squirted into the camel's face around the nostrils, through which they enter and then make their way down the pharynx, where the rest of their development occurs. When the larvae are just about to pupate they move up the windpipe, thus irritating the camel enough to bring about a violent sneezing and coughing fit resulting in phlegm heaving with prepupating maggots.

We have been acutely aware of these curious creatures for thousands of years thanks to us having a taste for animal flesh too (albeit thankfully now mostly cooked). Aristotle wrote about one species, a deer bot fly from the genus *Cephenemyia*, over 2,300 years ago. He stated that 'these creatures are as large as the largest grub' – grub being a term given to beetle larvae. Probably the largest grub he would have met is that of the European rhinoceros beetle, which can reach up to 10 cm (4 in) in length! Can you imagine something that long in your nasal cavity? I think it is safe to assume that when he used the term grub he was probably comparing this maggot with other, much smaller insect larvae.

Another amazing feat these creatures are associated with was recorded by Charles Townsend, an American entomologist, in 1927. He claimed to have observed a male deer bot fly, *Cephenemyia pratti*, achieving speeds of 400 yards per second – that is about 818 mph or 1,316 kmph. Indeed he later stated that fly collectors may sometimes need guns – a 22-calibre rifle to be precise – with its shells charged with sand to bring down the slightly slower females. Even a publication in *Science* in 1938 by Irving Langmuir, a Nobel-prize winner for chemistry no less, failed to demolish the claim from the popular press. He compared Townsend's observation of the fly with that of a Zeppelin (if they were of comparable size) to determine if there was any truth in this claim. Unsurprisingly, he found that there were indeed several reasons why these speeds could not have been reached, including the fact that, at this speed, the fly would be invisible to the human eye. He reasoned it was more likely to have been travelling at 25 mph or 42 kmph, which sadly wouldn't have produced the sonic boom science tells us that Townsend's fly would have (e.g. at 20°C its speed would need to be 761 mph). Townsend wrote a response to this rebuttal of his work offering explanations – Langmuir had made some bad calculations and, at the time that Townsend had proposed his own calculations, the speed of light was estimated to only be half of what it has since proven to be, and so he was amending the speed to between 400 and 818 mph. He still asserted his fly was invisible at top speed and 'leaves a visible blur in the air only when it suddenly decelerates to veer off and thus avoid collision with the observer'. Harold Oldroyd later wrote about this observation in his book *The Natural History of Flies* stating 'the original observation was made without instruments and was just a wild guess'. He was obviously not impressed with Townsend's sloppy observations. This incredible record was written about in *The New York Times* and subsequently the deer bot fly was added to *The Guinness Book of Records* for being the fastest creature on the planet. Later measurements of the speed of this species recorded 25 mph or 40 kmph, which is fast but not as fast as a male of the species of horse fly, *Hybomitra hinei*, which achieved speeds of over 90 mph (146 kmph) in pursuit of a female. That was one determined male.

Nasal bot flies generally favour large placental mammals but there is one species that is native to the marsupial red kangaroos of Australia. *Trachyomyia macropi*, the kangaroo bot fly, was first described in 1913 and scientists have been keeping an eye on it ever since. In the 2014 *Macropod Husbandry, Healthcare and Medicinals* book by kangaroo expert Lynda Staker, a case study was presented of a 10-month-old joey, named Nellie who was continuously snuffling. At first it was suspected that Nellie had a viral infection, but even after treatment she was still snuffling. Her handlers decided to give her, under anaesthetic, a nasal wash to flush out any fungal spores that they suspected might be causing these symptoms. Instead of seeing minute fungal spores, the vets witnessed the emergence of a large bot fly larva. Over the next seven days, and a course of the antiparasitic medicine ivermectin, another four came out! Poor Nellie. Though, in fact, she was lucky. Severe infestations have resulted in the deaths of their host due to bronchitis as many of the final instar stages get into the lungs and irritate the membranes. The host can also suffer from secondary bacterial infections brought about by tissue damage.

Another, slightly more famous Nellie, was Nellie the elephant. Now elephants suffer from infestations of bot flies too, but these are not, unlike the previous species, located around the nasal cavity – the bots would have a long way to travel if so! These species are from the subfamily Gastrophilinae, which, as the name suggests, develop in the stomach of the hosts. The larvae enter and exit through the mouth but spend most of their time developing in the stomach. It's not just modern-day elephants that have been affected by bot flies, as in 1973 the extinct species of the bot fly, *Cobboldia russanovi*, was found in the preserved remains of a mammoth's stomach.

As well as the mega-beasts, stomach bots are found in horses and rhinos. One of the most endangered species of animals on the planet may indeed be the rhino bot fly, *Gyrostigma rhinocerontis*, as not only do these flies have a very endangered host but they also have to locate each other to mate! This challenge is reflected in the very small number of specimens found in museum collections around the world. Few are from wild rhinos – that may be a slightly challenging task for any dipterist – but instead are donated from zoo animals. Indeed the original rhino bot fly was described from the stomach of a presumably dead rhino. I don't see the conservation societies leading an international campaign to save this little creature though!

The rhino bot fly is an impressive creature both in its adult and larval forms, with the adult being the largest fly known in Africa. We have a jar of their larvae at the Museum that we show to scientists researching the collections as they are some of the largest larvae of all the flies. Until 2012, it had been assumed that it was first described by Frederick Hope in 1840. However, Neal Evenhuis, an eminent American dipterist, then discovered that in fact the first description was by the Natural History Museum's very own founder, Richard Owen, in a catalogue of specimens housed in the collection of the Royal College of Surgeons, London, where he used to work, and it was published 10 years before Hope's species' description.

Most adult bot flies, despite the stomach-churning lifestyles of their larvae, are adorably fluffy, and they have very small, vestigial mouthparts as the adults don't feed. The larvae of both nasal and stomach bot flies are huge but, relative to the size of their hosts, tiny. That is not the case for the Cuterebrinae subfamily, which includes the New World skin bot flies. Again, these species have incredibly striking adults with massive eyes and tiny mouthparts, features I always feel make them appear slightly shocked. The large eyes aren't just to look pretty though, they need them to locate often very fast or very cryptic hosts on which to lay larvae. The tree squirrel bot fly or the American emasculating bot fly, *Cuterebra emasculator*, is a typical member of this subfamily and the name indicates that maybe the sympathies of the taxonomist who described this species

The exceptionally rare rhino bot fly *Gyrostigma rhinocerontis*, pupae and adult.

A deceased rodent and the pupa of a rodent bot fly, *Cuterebra fontinella*.

lay with the fly's hosts, as the squirrels often have large numbers of bots protruding from around their body.

The smaller the mammal host the larger the larvae seem in comparison. The rodent bot fly for instance, *Cuterebra fontinella*, seeks out mice, chipmunks and other similar small-sized rodents. And the maggots are often bigger than the host's offspring.

Not to be outdone in the size of their eyes, the big-headed flies of the family Pipunculidae are quite spectacular, with their massive eyes making up most of their heads, which are connected by the smallest of necks to their bodies. Their heads are disproportionately large in comparison to their bodies and so micro-pinning them successfully with their head still attached is a minor miracle! Both males and females have these incredible eyes and it gives them a massive edge when it comes to mating and finding a host. A fellow dipterist friend of mine often muses about how these flies copulate mid-air and successfully live to fight another day. Their hovering ability, like those of their closely-related relatives, the hover flies, is exceptional. Pipunculids can hover in very confined spaces, such as amongst the dense, low vegetation where their larval hosts, the Auchenorrhyncha, are found.

Auchenorrhyncha is one of the suborders of true bugs (all bugs are insects – not all insects are bugs) that include the hoppers, cicadas and spittlebugs. These little bugs keep very close to the vegetation that they're feeding on and are often well camouflaged. The Pipunculids need to spend a good deal of time seeking out their hosts in what can be very shaded habitats. Interestingly, some species pluck their hosts from the vegetation before laying eggs in them mid-air and then placing them back on the plant. The females slice and insert their eggs into either adults or immature hosts. It was thought for a long time that bugs were the only ones parasitized, but in 2005 David Koenig and Chen Young, two American entomologists with an interest in crane flies, were out collecting these species when they discovered parasitized crane flies and reared out *Nephrocerus zetterstedt* specimens. This was unusual, not just because it was the first time non-bug hosts had been recorded but also because it was the only other incidence of flies parasitizing flies outside of the conopid, tachinid and house fly families.

Some scuttle flies also have body-piercing ovipositors. Flies in the genera *Pseudacteon* and *Apocephalus,* and further species from the subfamily Metopininae, have blade-like ovipositors that enable them to pierce through the backs of ants and then insert their eggs. The larvae crawl through the body cavity into the head where they consume the contents for the next two to three weeks. Strangely, the ants don't appear to be affected at all during this period and carry on as if nothing unusual is happening. But this gradual consumption eventually leads to the head breaking away from the body, effectively decapitation. Some species of scuttle fly also release a digestive enzyme at the final stage that dissolves the neck, resulting in the head falling off. Either way the larvae then pupate in a perfectly protected cavity for a further couple of weeks until they emerge as adults to strike fear once more into the ants. Some species have even modified their mouthparts to resemble a saw. Brian Brown, an American dipterist specializing in these flies, managed to film these odd creatures in action. These phorids are from the genus *Dohrniphora* and are normally regarded as scavengers. It seems that a few of these species were not informed about this fact though, and instead are parasitic, targeting injured female trap jaw ants, *Odontomachus*. These ants have the fastest predatory appendages of any animal, snapping their jaws shut at average speeds of up to 130 m/s. The female fly darts around the injured ant, dashing in now and again to check whether it is still capable of inflicting any damage. Once she has determined that the ant is no longer a danger she starts sawing through the neck of the dying female. Brown describes them cutting through the tissue of the gut tract, the nerve cord and the intersegmental membrane (the part between the head and the thorax). He actually filmed the tiny fly – the head of the ant alone is bigger than the fly – sawing and tugging at the head until eventually it came loose and the successful fly ran off with it. This may seem like a particularly nasty death for the poor ants but these parasitoid flies are a good thing for humans. Fire ants were accidentally introduced into the USA in the 1930s from South America and they have been wreaking havoc ever since. The aggressive little ants

A female phorid fly *Dohrniphora* sp. attacking a dying trap jaw ant, *Odontomachus* sp.

202

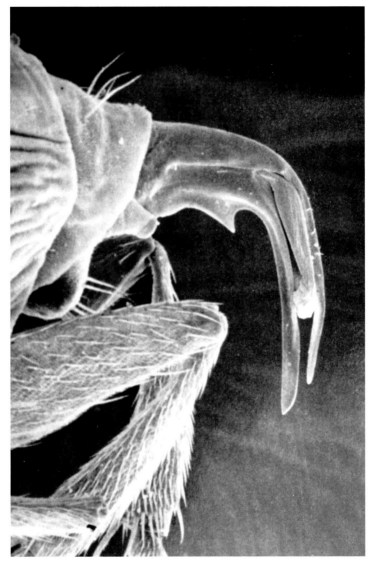

The hooked ovipositor of *Pseudactus curvatus*, used to prize apart the segments of an ant.

like to live near human habitation and so we often come into direct contact with them, resulting in notoriously painful stings and bites, a glorious double-ended attack. Luckily for us though, the phorids are here to help us. Classically, ant-decapitating phorid females hover above the ant and pick the right moment to dart down and lay an egg in the back of the highly aware ant, which desperately tries to evade her.

Although great at killing ants, some species of phorids have turned their attention to honey bees. Up to 13 *Apocephalus borealis* larvae have been found in a single dead bee. And to make matters worse these flies tested positive for – and due to high densities are serious vectors of – deformed wing virus and *Nosema ceranae*, both of which are important pathogens of bees. It seems that, finally, this fly family's ecological adaptability has led them to become problematic for humans. This fly causes the original hosts – bumblebees and paper wasps as well as the honey bees – to leave their nests at night. These hosts have now received the moniker of 'zombees'!

Another group of parasitoids are found in a cluster of closely-related families: the small-headed flies (Acroceridae), the tangle-veined flies (Nemestrinidae) and the bee flies (Bombyliidae). As adults they are all adorable, with some of them looking like little powder puffs, but it is their parasitoid offspring, which have adapted to living in some extreme environments, that grab all the headlines.

The small-headed flies are also called spider-killing flies, which gives a nod to what creatures the larvae consume! The adult flies have massive, humped thoraxes with tiny heads, and so have also affectionately gained the name of the hunchback flies. These are rarely seen as adults and most of them come to the Natural History Museum collection by way of a disappointed spider specialist. The larvae are indeed spider parasitoids and these larvae, as with the tangle-veined flies and the bee flies, undergo hypermetamorphosis. This means that there is not just one larval form in these families but two distinctly different morphological and behavioural larval stages.

The first instar stage found in all three families is called a planidium and, as with the tachinids we discussed in Chapter 3, this is a very active stage. The planidia in flies differ from the planidia found in beetles and wasps insofar as they do not possess legs. You may think this would hinder them in locating their rather active hosts, but there are some helpful adaptations that enable them to overcome their physical handicaps. *Ogcodes*, the largest genus of small-headed flies, parasitize mostly wolf spiders but will attack a range of other spiders including the very fast jumping spiders. To have at least a remote chance of success, the female fly lays an extraordinary

Possibly the cutest animal on the planet – the grey bee fly, *Anastoechus melanohalteralis*. Cute adults, but their offspring are parasitic.

number of eggs. Various accounts include figures as high as 3,000 being laid over a four-hour period or 5,000 eggs being laid over a 10-day period, often in aggregated clumps. Females at maximum egg-laying productivity have been reported to achieve rates of one egg laid every five seconds. Many perish at this stage through dessication or being munched on. If they survive, however, the freshly emerged larvae immediately stand erect in wait for their hosts to pass by. They have a hooked sucker that enables them to attach to a surface and some long hairs at the rear of their body for support, and they can swing themselves around. The larvae have mouthparts armed with a piercer and a pair of hooks to enable them to grab onto a passing host and then start crawling up their unfortunate victim.

If they are left waiting, with no suitable hosts passing by, the larvae can help their situation by 'walking' to a more favourable place. Firstly they can 'leap' 5 or 6 mm (¼ in) at a time by bending their body over and pinging themselves across spaces. They have been observed to move in sync with their hosts enabling them to approach undetected – they only move when the spider moves. Their other way of moving is inch crawling, a method used by many caterpillars where the rear legs are brought up to the front legs before the front legs are moved forward, and this enables them to shuffle across the silk strands of webs. There is no immediate panic to locate a host as they can survive, suckered to the ground, for up to six days, wiggling in wait. If the little larva's luck is in and it comes into contact with a host, most species crawl up the spider's leg and enter through its abdomen. Some larvae have been observed crawling in through the segments of the leg but this is not thought to be a common entry point. Once inside, the first and second instar larvae can take six to nine months to develop, feeding on the host but not with any serious intention. They may even enter a period of dormancy if the spider is very young, and wait until their host is mature, which may be a number of years. The third instar larvae make up for the inactivity of the first two stages. They are voracious feeders, eating all but the vital organs, which are consumed just before the larvae is ready to pupate, thus keeping the spider alive and the food fresh for as long as possible.

The larva emerges from the host and pupates outside of it. Pupae of the small-headed flies partially resemble the adults that they are morphing into, in that they have a distinguishable small head region. Some are solitary while others are gregarious. Soren Toft, a Danish spider expert, found that if there was only one larva in the wolf spider, *Pardosa prativaga*, then the spiders were able to suppress the infestation. More than one larva – and they recorded up to eight in an individual host – resulted in the spider no longer being able to cope and there was a negative impact on its lifespan. In small-headed flies the pupal stage can be very short, two to three days or up to a month, but with some species of their cousins, the bee flies, this stage can be between three and four years. The final adult stage for the small-headed flies is also a short one, with species varying between three days to a month, and they spend their time dispersing and copulating.

Bee flies parasitize many different species but do have a particular fondness for solitary bees. With these species the larvae pupate in the cells of the bee burrow alongside the developing bee larvae. Just before hatching, the fly pupae leave the cells and wriggle up to the entrance to the burrow. They have spines along their bodies which provide the traction necessary for them to do this, and the pupae to leave the nests. Once more the flies challenge the preconception that the pupal stage is a dormant one.

In some cases, as the adults emerge away from their larval hosts, we are left in the dark as to what their hosts were. This is not the case with the next group of flies as they're all ectoparasites, meaning they live externally on their hosts. Unusually for flies it is the adult stage of the ectoparasites that feeds on the hosts – the larvae are fed directly by the mothers. The most notable of ectoparasites in flies belong to the bat flies (Nycteribiidae and Streblidae) and louse flies (Hippoboscidae) as they are some of the oddest-looking species, with many lacking the features we consider diagnostic for flies: the wings. I have undertaken fieldwork in the Caribbean and was lucky enough to work alongside bat specialists. At night they would catch bats and, while they were jotting down the details of bat species, weight etc., we would remove any flies that were crawling over them. Seeing these bizarre wingless creatures running around on the underside and backs of bats is truly an odd but fascinating experience – these things move around like drunk spiders. My own interest in these splendidly weird-looking flies was sparked by one of my colleagues at the Museum who, a while back, left a rather strange creature on my desk that she had removed from a pinned mantid in the collection. At first she thought it might have just been some 'crud' (technical term for debris), but on closer inspection it turned out to be a bat louse.

Usually both the families Nycteribiidae and Streblidae are called bat flies. In the nycteribiids all the adults are wingless but they have retained their halteres or balancing organs, which enable them to move around on the host. Their wing muscles have completely atrophied (dissolved) and so they have an incredibly small thorax. The head and legs basically just look like extensions of the thorax and the head sticks out of the top of the thorax after the first pair of legs rather than in the more usual position in front of the legs. Their tarsal segments have been completely bent round and are furnished with huge claws, which help them cling onto the bat's fur. It is really difficult to work out which way is up as the head doesn't look like a head either. Instead their heads are small hairy ovoid structures, often without eyes. These species are really adapted to their ectoparasitic lifestyle with all these bodily modifications.

Adults average only 5 mm (¼ in) in length and so move about with ease through the bat's fur. These tiny adults live exclusively on bat blood, and both males and females can consume their own body

The enlarged tarsal claws of the bat lice fly, *Nycteribiid* sp., which enables it to hold on to bat fur. The head is very small and often there are no eyes.

weight every five days. These species do not count as sanguivores or bloodsuckers because they live and feed permanently on their hosts unlike the former, and only leave their hosts to give birth.

Some species of the Streblidae have taken this body modification to extremes, with the female becoming nothing more than a sac, or rather a glorified womb. Within the genus *Ascodipteron* the female initially starts out resembling an average-looking fly, but on finding a host she embeds herself in it. She then nips off her wings and her legs and undergoes a post-imagina metamorphosis – she goes through a second period of change to resemble once more a maggot!

Another wonderful little ectoparasitic group is the bee lice, Braulidae. As with the nycteribiids, these are all wingless and, as their name suggests, live on bees. What makes these flies even more unusual is that they don't have either the simple eyes (ocelli) or a scutellum (lobe-like structure at the posterior end of the thorax), but more fundamentally they don't have halteres. Halteres are essential as balancing organs in flight and as such are not so important for the flightless species. But this is the only group to lose them completely. Adult bee lice share an intimate relationship with bees, and sadly because of this they have been wiped out, sometimes intentionally sometimes not, in many countries including the UK. These flies are not parasites in the true sense of the word but instead are kleptoparasites – they steal food from their hosts! They are generally

The wonderful bee louse fly, *Braula coeca*, that steals food from its host, the bee, and which is now extinct in the UK.

found on queen bees and, as the queens are fed the most, the fly is guaranteed a good supply of food. Their presence was initially not seen as too problematic, as normally only a few were found per host (although numbers of up to 180 have been recorded). Their demise came about when beekeepers discovered that the varroa mite was having a detrimental effect on honey bee colonies. To counteract this beekeepers started using miticides in the hives which also had a disastrous effect on fly populations. The mites generally live on the developing bee drones, which would have, originally, been removed from the hive and were often traditionally feasted on by the beekeeper. I have eaten these and can testify to how tasty they are. But now that drone larvae are left in there, the mite is able to proliferate and it appears that in the UK at least one of our most adorable species has gone. Bee lice have not been seen in years but, oddly, I have not as yet heard of a braulid reintroduction programme from any of the wildlife groups – once more the charismatic mini-fauna lose out.

Parasites are often looked at in horror, and to be fair some do have rather gruesome survival strategies, but they are fascinating in both their life histories and their morphology. They suffer greatly from bad press, from people who sadly cannot see what truly remarkable species they are. Their bodies are some of the most extreme in terms of modifications from the basic plan, which of course leaves even the most experienced of dipterists with feelings of childlike pleasure.

CHAPTER 10

The sanguivores

I behold you stand,

For a second enspasmed in oblivion,

Obscenely ecstasied,

Sucking live blood,

My blood.

D. H. Lawrence, The Mosquito

O UR LAST GROUP of flies contains the one that probably annoys humans the most - the mosquito. And a question I am often asked, in a less than pleasant tone I may add, is why do we have mosquitoes at all. What's the point of them? Most people hate these vampiric, albeit delicate little creatures. This is understandable, as many of the flies that suck blood – the sanguivores or haematophagous feeders – can act as vectors or carriers of some pretty nasty diseases. However, it's rarely the bloodsucking flies themselves that cause death. Instead it's the parasites, viruses and bacteria transported in them, the pathogens, that are the killers. Rather than focusing on the diseases spread by these flies I am going to concentrate on the flies themselves and how they have adapted to their sanguivorous life cycle.

The holoptic eyes of the black fly, *Simulium damnosum*, showing the two different sized ommatidia.

In most of the sanguivorous species it is the females that are the bloodsuckers – the males are mostly vegetarians, feeding on nectar or other plant products. Females may also supplement their diet by feeding on nectar, but blood is often the critical requirement to ensure egg development. Blood is incredibly protein-rich, composed almost entirely of red blood cells which, in mammals, predominantly consist of the protein haemoglobin (96%). The other 4% is composed of lipids – fats, sugars and other essential components such as nucleotides. Protein is essential for growth and development but can be sparse in nature. If you're a grass-nibbler you need to eat a lot of grass as grasses have a protein content of, at maximum, 30%. There may be a lot of vegetation around to munch on but it takes time and effort to both source large amounts and then digest the complex structures and absorb the nutrients. You can eat smaller quantities of food if you consume animals – the amount of protein in insects, ranges from 13 to 77% depending on the insect – and so you generally need to eat less. Bloodsucking is relatively rare in insects but flies once again seem to have adapted and thrived on this rather extreme diet, with at least 14 families containing sanguivorous species. Why not just go to the best primary protein source, blood?

Bloodsucking is thought to have had two ancestral routes. The first is where predatory flies evolved to specialize in bloodfeeding. This happened with the snipe flies of the family Rhagionidae, most of which are predatory, but a few switched to bloodfeeding. The second ancestral route evolved through the close associations many flies have with other species. For example, many species of fly living in bird nests or bat roosts originally fed on the waste or by-products of the vertebrates, but have now evolved to feed directly on the vertebrates themselves. This has two advantages: firstly they still benefit from the shelter of the host and secondly it separates the food source of the larvae and the adults.

All flies, irrespective of their feeding habits, have sucking mouthparts which are often elongated, and one or more suction pumps to aid fluid uptake. In the bloodsucking groups such as mosquitoes, biting midges and horse flies, the mandibles of the females have elongated their proboscises and these are often sharpened and/or adorned with tiny rows of teeth, resembling minute saws.

It's these protrusions that enable the fly to pierce through the skin of mammals, birds and frogs and so feast on the fluids beneath. Just about anyone who has been bitten by a mosquito or biting midge will testify as to how annoying, painful and irritating such bites can be. This reaction is mostly due to the different chemicals in the fly's saliva, but that does not detract from the fact that there are some vicious slicers out there. Flies obtain blood by two methods: either they pierce directly through the skin to a vein (as with mosquitoes), or they slice through the skin and lap up the pooling blood (as with female horse flies, which have some of the most impressive mouthparts – hardened stabbing devices with sharpened blades along the edges). Female horse flies are not totally vampiric and will spend time nectar-feeding before they obtain a blood meal necessary for egg development. They are also the only flies that combine two functional mouthparts into one. We have come across the *Philoliche* species of horse flies found in South Africa, in Chapter 2, where the females, though important pollinators, require a blood meal in order for their eggs to develop. The female uses the short mouthparts located closer to the head for piercing the skin and sucking up the pooling blood, and the very long mouthparts for nectar-feeding from the long-tubed plants, so she has two functional food canals. The saliva of these flies, as with all sanguivores, contains anti-coagulants that slow the clotting of blood, therefore enabling them to feed, if undisturbed, for longer – once the blood clots they are no longer able to suck it up their food canals.

Horse flies are active and persistent biters and frequently come into contact with humans. They have many common names including deer flies, clegs and gadflies. Gadfly is, amusingly, now used in common language to describe an annoying person who is persistent in their criticism – in the same way as the fly is persistent in pestering for blood. Their reputation as vile creatures is compounded in Central and South America where they not only attack humans but are one of the transporters of the human bot fly, *Dermatobia hominis.* The female lays packets of eggs onto the back of the intermediate host, which are then cemented in place. The heat of the final host stimulates the larva to emerge and penetrate the skin. Just think, you can be sliced apart and sucked on by the horse fly, and then further feasted on by the baby bot! Mosquitoes and house flies are more common carriers of the bot fly, but horse flies are almost silent in flight and so make excellent carriers as they can approach the host without attracting attention to themselves, unlike the whiney mosquitoes.

More than 4,400 species of horse fly have been described to date. They are one of the larger fly families and are all chunky, robust creatures, often with the most spectacular bands, squares, triangles, circles and wiggles on their eyes. These patterns are all the result of cornea colour filters, the outer layer of their eye facets. These reflect different wavelengths of light and so their eyes reflect different colours when observed from different angles. Sadly, when the fly dies its body dries out, including the rather moist eyes which collapse back into the head. You can rehydrate them, even years after the specimens have died, and the patterns reappear for a short time, which can be useful for identifying species. These flies also have different feeding preferences across the genera. The genus *Haematopota,* commonly referred to as the clegs, attack at waist height, while the genus *Chrysops*, the deer flies, generally aim for our head.

The deer flies, *Haematopota pluvialis,* with spotted wings and banded eyes and *Chrysops caecutiens,* with spotted eyes and banded wings.

Both of these formidable females can remove a fair amount of blood during a feeding frenzy on a sunny summer morning or late warm afternoon, which is when they are most likely to attack. It is not uncommon to see vast numbers feeding on one poor cow. Cornelius B Philip, an American researcher working on horse flies, noted that 20 to 30 of these flies, feeding for six hours, can take up to 100 ml (3½ oz) of blood. Draining mammals of large quantities of blood is not limited to just horse flies – there are many families that specialize in this behaviour. But the most famous of the bloodsuckers, the mosquitoes, take very little blood individually, even though they often cluster in large numbers and so appear to be the most voracious.

In the house fly family there are some formidable bloodfeeders in the genus *Stomoxys*, the stable flies. No longer satisfied with a decomposing diet that is more typical of the adults of this family, species in this genus have modified their mouthparts for a sanguivorous diet, resulting in some of the most painful bites given by an insect. I had a discussion with my colleagues recently over which flies they thought inflicted the most painful bites, and the stable flies were high up on the list. As with most biters, at the end of the long, slightly thickened proboscis there are some large, well-developed teeth that are able to shred through the skin of their victims.

The preferred hosts for stable flies are large mammals including horses, donkeys and mules, but they also regularly attack humans. Females have been observed feeding for 10 to 15 minutes on the unlucky hosts if not disturbed. I have been lucky enough to go collecting for these beasts with a group of international entomologists in the Coptic Orthodox Church forests of the Ethiopian Highlands. Many of the livestock that graze the countryside seek shelter in these forests, but this comes at a price. Although offering shelter from the Sun, there are vast numbers of

The ubiquitous stable fly, *Stomoxys calcitrans*, of the house fly family. Its bite lives up to its fierce reputation.

Stomoxys waiting for their dinners in the cool climate of the forest, as was demonstrated by the huge densities in my samples. Unusually for bloodfeeders, the males also have a haematophagous diet, and both sexes have a voracious appetite, feeding two to three times a day, and they can travel long distances – up to 25 miles (40 km) in one day – to satiate themselves.

The elongated and hardened proboscis of the stable fly, *Stomoxys calcitrans.*

As well as being tenacious, stable flies are also quite clever. William Thorpe, a zoologist at Cambridge University, wrote about having the great fortune to spend a prolonged period of time during 1939 working on bugs at a research station in Amani, nearly 1 km (1½ mile) up the East Usambara Mountains in Tanzania. Whilst he was there he became

The end of the proboscis of the stable fly, *Stomoxys calcitrans*, has some very large well-developed teeth for shredding its prey.

obsessed with the local populations of army ant, *Dorylus molestus*, and observed them having their prey snatched from their jaws by the blow fly *Bengalia depressa*. These flies are attracted to disturbed ant colonies where the adults snatch the food and pupae being carried by the ants. However it wasn't the ants or the blow fly that caught his attention. On an evening ramble with his wife, he spotted a fly hovering above a column of returning ants, and noted that a creamy-white object, which he suspected was a larva, 'extruded from the abdomen' of the fly and was dropped in front of an unburdened army ant. The ant dutifully picked it up and carried it into the nest. Thorpe caught the fly, but lamented that he also caught some ants along with it which then proceeded to attack and shred the fly. Although the fly had been mutilated, he was still able to pin it, and the specimen was eventually identified by entomologist Helmut van Emden as *Stomoxys ochrosoma*, the adults of which have been observed feeding on mules and baby buffaloes. We don't know what the larvae were feeding on, whether the ants or the ants' waste, but once more it highlights the variety in diet between the adult and the larvae.

Haematobia and *Haematobosca*, as their names suggest, are also bloodfeeders, both genera of the house flies. Commonly called horn flies and buffalo flies respectively, *Haematobia* are of similar appearance and behaviour to the stable flies, whilst *Haematobosca* species lack the thickened proboscis. Horn flies were originally only found in Europe, the most annoying one having the appropriate name of *Haematobia irritans*. Sadly this species didn't stay put and, after being accidently introduced to North America in 1889, it spread over the two continents. It had a major impact on the cattle industry because these flies cluster in huge densities. Figures from the USA alone estimate that these flies cost the economy $700m to $1 billion annually in cattle losses and control measures.

Horn flies, *Haematobia irritans*, clustered in a feeding orgy on a cow. Originally only found in Europe it has spread to North America and impacted the cattle industry.

These horn flies are formidable little trackers and can smell their intended victims from distances of 7–10 miles (11–15 km). The expansive rangelands across the USA present no problem for these little travellers. Again, both sexes are bloodfeeders, but the females are more aggressive and often ingest twice as much as the males – which means they can sometimes feed up to 40 times a day! The females are very particular about where they lay their eggs, requiring the freshest of dung, sometimes so fresh that the cow hasn't even finished defecating before they start laying.

The moose fly, *Haematobosca alcis*, mainly attacks, unsurprisingly, moose. It is a house fly and not to be confused with the moose flies mentioned in Chapter 6 – a confusion that wouldn't happen if we just used Latin names. It was once referred to as a lost species as, for 30 years after it was originally sampled and described, it wasn't seen again. Even now it is rarely caught in most samples which is odd as a survey in 1974 in Yellowstone Park in the USA found that as many as 500 individuals were seen feeding on a single moose. Remarkably the moose appeared quite calm about the situation.

No adults in the closely-related blow fly family have yet been found to be sanguivorous, but there are many species whose larvae are, including one that has a very intimate relationship with humans. The Congo floor maggot, *Auchmeromyia senegalensis*, is found in sub-Saharan Africa and Cape Verde, off the coast of West Africa. The larvae live in the cracks and crevices of animal and human dwellings, leaving these at night to feed on sleeping inhabitants, thus earning themselves the nickname of vampire maggots. All five species in the *Auchmeromyia* genus are bloodsuckers but the others don't feed on humans, instead preferring wart hogs and aardvarks.

Tsetse flies of the family Glossinidae are not subtle feeders and have an exceptionally painful bite. They have what has been appropriately described as a bayonet for a proboscis, bulb-shaped at one end and which both sexes use to stab their victims. The adults can feed for up to 20 minutes then drop off and hide, repeating this nightly. Their proboscis is similar to that of the stable flies, but the adults themselves are slightly flatter in shape and so more closely resemble their cousins, the louse flies. Only 23 species found within one genus, *Glossina,* comprise this family but even with limited numbers they get themselves into lots of trouble. It is the only genus in all the flies to have a hatchet-shaped discal cell on its wings, which resembles a machete and which suits their aggressive character. As well as inflicting harm on their hosts, they are also key vectors for some diseases, the worst of which causes sleeping sickness in humans.

Tsetse flies are in the superfamily Hippoboscoidea and they, like other families within this group, have developed a method of reproduction called adenotrophic or gland-fed viviparity, live birth. The female only produces one egg at a time, which she retains inside her, and this egg contains enough yolk for its development into a larva. Tsetse flies mate just once, with males ejecting a sperm package called a spermatophore into the female. The female stores the sperm in a paired organ called a spermathecae from which she releases the stored sperm periodically. The egg and sperm meet at the head end of the ovary and eventually the fertilized egg passes to the uterus to develop. The larva breaks out of the egg case by using a temporary egg tooth located on its head. Its emergence is further assisted by the egg case coming to rest on a highly invaginated, or inwardly folding, sticky surface of a specific area of the uterus, the choriothete. The egg attaches firmly to this, enabling the larva to push against the uterus wall and wiggle out. All three larval stages are internal in the adult. The female converts the incredibly rich food source, blood, into a milky substance which the larva feeds on via an internal lactating gland. The final instar stage is a very odd-looking creature with what appear to be ears arising from its rear. But these are not used for hearing – they are its respiratory organs, called polyneustic lobes. The larva's diet is so incredibly nutritious that on emergence it weighs as much as its mother and occupies most of her abdomen.

Most bloodsuckers are found in the suborder Nematocera. Among them, the Corethrellidae, or frog midges, specialize in feeding on frog's blood. They are superficially similar in appearance and behaviour to mosquitoes and were originally placed in this family. But they differ in wing venation and are very small in comparison to the mozzies, with many being less than 2.5 mm (⅛ in) in length. These strange and wonderful flies number just 66 species in two genera and are found globally, generally within the tropical belt (from 50° north and south of the equator).

Frog midges feeding from blood in the nose of a hapless male frog. Only the females feed on blood.

In frog midges, as with all nematocerans, only the females feed on blood and they have two slightly different ways of approaching the frogs to get their meals. If the male frog is calling intermittently, with periods of silence, the approaching female midge proceeds with short bursts of flight, dropping to the ground when the calling pauses. With the resumption of calling the female once again takes to the wing and uses the frog's singing to mask her approach. When she is within 20 cm (8 in) of the host she stops altogether, and then sneaks silently up to the frog and climbs on to it to feed. If the frogs have a continuous call she flies in, bobbing up and down as she approaches, striking all surfaces until she reaches the right one – the frog. The bloodsucking adults, as with many bloodsucking flies, have larvae that we might think of as beneficial – they feed on the larvae of species that are harmful to us and our animals. *Corethrella* larvae are known to feed on mosquitoes and nematodes and have even developed prehensile antennae and some pro-legs (fake legs) to enable them to grab at their prey. Trials are underway to study the effectiveness of using them as a biological control agent for the Asian tiger mosquito, *Aedes albopictus*, a prolific species that is also a vector of Dengue fever, Japanese encephalitis and Zika.

The *Deinocerites* genus of mosquitoes comprises 18 species, all found living in the burrows of crabs in North, Central and South America. The crabs are not harmed as the female mosquitoes feed solely on birds, particularly the storks found in the associated wetlands. The crab burrows are used for both the larvae and adult stages, but just as shelter. The adult male emerges earlier than the female and waits patiently and observantly for her to emerge. The adult males of one species, *Deinocerites cancer*, have greatly enlarged tarsal claws on their first pair of legs and also have exceptionally long antennae. These long antennae differ from most other mosquitoes in that they have fewer fibrillae – the thread-like hairs associated with hearing – but more *sensilla basiconica* (peg-like structures)

The wonderfully long antennae and tarsal claws of the *Deinocerites cancer* mosquito.

The sunning female mosquito, *Sabethes tarsopus*, showing off her fabulous leg warmers.

and *sensilla camponiforma*, which are instead associated with odour perception and movement respectively.

Mosquitoes are rather sensitive creatures. They are able not only to identify members of their own species through differences in wing vibrations, but the males can harmonize themselves with prospective mates. Instead of listening for the females, the male smells and feels for her. Although these are unusual modifications, they play a key part in locating the aquatic female pupa and guarding it from other suitors. We are still not sure how the male is able to tell whether the pupa is male or female whilst resting on the surface of the water body, but he does, and then he guards it (if female). As she emerges, he seizes her air trumpets – the breathing tubes at the head end of the pupa, pulling them apart and so helping her out of her pupal case. Once copulation has been initiated males have been observed rapidly vibrating their wings for up to half an hour until finished. This is a very long time for your average mosquito, where mating usually takes less than a minute. But by having enlarged claws, as well as genital claspers, the male is able to successfully cling on during this period.

Arguably the most attractive of all mosquitoes is the genus *Sabethes*, which live in forests. Both male and female *Sabethes* are brilliantly coloured, with metallic scales and sometimes paddles (feathers) on their middle legs. Female adornment is unusual in any species as usually it is the males which are tasked with trying to attract females. Mosquitoes are no exception, with males from many species using paddles in courtship displays (as discussed in Chapter 8). So why should these females have paddles as well? Well the truth is we don't as yet know. In experiments. paddleless males were still as successful in the mating game as paddled males, but paddleless females were rarely approached, suggesting either that females just weren't interested in paddles when it comes to choosing males or that the females were the active sex in courtship behaviour and that indeed it was her adornments that determined mating success. Paddleless males and females had no problems in flying or in oviposition so it was not due to any physical difficulties that the females weren't successful. The males just didn't like them not being there. No-one is sure why this is the case, or what the paddles are for, but most agree they are really rather attractive.

Another pretty little species is the eastern saltmarsh mosquito, *Aedes sollicitans*, which is found along the east coast of North America and the Caribbean. She is opportunistic in her hunting and has been noted to travel over 40 km (25 miles) to find a host. This species has been incriminated as a carrier of both eastern and

The eastern saltmarsh mosquito, *Aedes sollicitans,* with its stripey legs, is opportunistic in hunting and can travel over 40 km (25 miles) to find a host.

Venezuelan equine encephalitis and dog heartworm, and it has also been blamed for taking so much blood from cattle that the animals die, a process referred to as exsanguination.

Back in 1980 Hurricane Allen hit the Caribbean, Mexico and the southern United States, resulting in thousands of acres of southern Texas ranch land being flooded. And with the rains came the mosquitoes. As the water receded 15 cows were found dead from what appeared to be anaemia. Dr Bruce Abbitt, who was then working at the Texas A&M Veterinary Medical Diagnostic Laboratory as a veterinary pathologist, worked out that the cause of death was in fact due to extreme blood loss from mosquito feeding. He estimated that the number of bites necessary to kill a cow through blood loss would be some 3.8 million bites – the mosquitoes would remove half the blood from an average cow, approximately 20 litres (35 pints), for death to occur – an extraordinary figure. In comparison, humans only need 2 l (3½ pints) to be removed before they start getting into dire straits.

A remarkable fact worth mentioning here about bloodfeeding females is that they have learnt how to cope with the huge hike – 10°C (50°F) – in their body temperatures that it causes. Mosquitoes exhibit heterothermy, where different parts of their body cope with differing temperatures. The hike in temperature may also lead to problems such as a fellow mosquito mistaking it for a host and trying to pierce it and feed from it. The *Anopheles* mosquito has learnt to deal with this by excreting a droplet of fluid from their anus whilst feeding (a process called prediuresis). Thanks to the evaporative cooling effect of these droplets, the females are able to limit thermal stress.

Black flies, the simuliids, can also occur in massive swarms and cause death through exsanguination. Veterinary and medical entomologist Claude Noirtin and colleagues working across Europe describe an outbreak during the spring of 1978 when 26 cows were killed by thousands of *Simulium ornatum*. This species is unusual in that it generally bites the belly of cows rather than the backs and the legs, which are attacked by other species. Noirtin and colleagues estimated that the dead animals had on average around 25,000 bites, and in one case the poor cow had over 55,000 bites. As with the mosquitoes, only the female *Simulium ornatum* are bloodfeeders, but both sexes were found in the swarms. These swarms may take the form of a cloud that clusters over your head and moves with you while you walk, like a giant copulatory rave. An interesting case in July 2015 was that of a Canadian woman with hundreds of bites on her ankles. She describes how she had not been aware of them when she was outside gardening, but when she went inside her husband noticed blood oozing from her skin. Interestingly the bites were in regular lines, a very uncommon biting pattern for this species. Advice that should always be heeded: when there is a swarm above you, and you are with friends, you should bend lower than them so the swarm moves onto one of them.

Black flies do not have the slim, delicate appearance typical of Nematocera, like mosquitoes and crane flies. Rather they are chunky, robust flies – picture a flying bulldog and you are on the right track. Even the characteristic thread-like antennae of this subfamily are missing as they have very compact antennal segments that resemble a tube of Maynards wine gums.

A European species of black fly, *Simulium posticatum,* came to be called the Blandford fly in the 1960s and 1970s because there were such large outbreaks of them around the town of Blandford Forum in Dorset, UK. This species is not restricted to the UK and is found across Europe, and it gives a very painful bite that swells and is incredibly itchy. By the 1980s thousands of people had been affected by this species in the UK and in a bid to control the problem the government sprayed a biological insecticide, *Bacillus thuringiensis israelensis* (Bti), onto the waterbeds where the larvae lived. A protein that is synthesized by this bacterium breaks down and perforates the larva gut, killing it in the process. This was a very successful campaign and reduced the populations dramatically to just 10 % of their original densities. There was another, very British remedy to these bites in the form of a beer. Initially called Blandford Fly (it is now Blandford Flyer) this ale was rumoured, I am presuming by the brewers themselves, to be a recipe for easing the itching and pain caused by these little flies. Drink enough and I'm sure it was.

A more typical Nematoceran-looking family are the biting midges of the Ceratopoginidae, a closely-related family to mosquitoes and also predominantly a bloodsucking group with slender bodies and long antennae. They attack both vertebrates and invertebrates – especially insects. I have already mentioned the genus *Forcipomyia* as pollinators of chocolate in Chapter 2, but many species are

Phoebis sennae caterpillar playing host to many adult *Forcipomyia eriophora* flies.

bloodsuckers of lacewings and their relatives (neuropterans), the caterpillars of sawflies, butterflies and moths, and spiders, crane flies, dragonflies and damselflies.

Originally it was thought that just one species of *Forcipomyia* was feeding on all these hosts, but now many species have been incriminated. These tiny flies are seen clinging to various body parts of the adult hosts including the wings of the flying insects and the abdomens of spiders, sucking the lymph from their veins. Arguably some of these should indeed be termed ectoparasites, as the females spend most of their lives attached to the hosts, so resembling mites in behaviour. When the females stop feeding prior to giving birth, they resemble mites in morphology too, being rather rotund.

One interesting biting midge, although a bloodsucker of vertebrates, actually gets its blood second-hand. *Culicoides anophelis*

The hunter becomes the hunted – the midge *Culicoides anophelis* feeding on an engorged *Anopheles* mosquito.

is a very small fly – smaller than 3 mm (⅛ in) – and its diminutive size helps the female's rather sneaky behaviour. They do not feed directly from the vertebrates but instead steal the blood meals from much larger mosquitoes. Described back in 1922 by the dipterist Frederick Edwards, female *Culicoides anophelis* were sampled from a group of *Anopheles* mosquitoes and initially were thought to be feeding on them. Eventually it was determined that the females weren't feeding directly on the mosquito but on the mosquito's last blood meal. They don't just plague that mosquito but have now been found on a variety of other mosquitoes. The midge lets the mosquitoes do all the risky stuff – getting up close to the large and dangerous vertebrates to take blood – and then she comes along later and helps herself to the proceeds. It's a bit like a flying fridge for the midge. Some of them are thought to be phoretic on the mosquitoes – using them as transporters as well.

Like them or hate them, sanguivores have had to adapt to a unique and often dangerous lifestyle. I can directly testify to this as we have, within the bowels of the Natural History Museum collection, a donated collection of horse flies. All of these are beautifully pinned and labelled, but every single one of them has been flattened, presumably because of the collection technique the donator had decided upon – they had gorged themselves on the wrong person!

The end

S O NOW YOU'VE MET the flies. That is by no means the end of the story for these fantastically useful, adaptable and diverse animals. I've so many more tales that didn't quite fit into any of the preceding chapters, but hopefully there have been enough to highlight the importance of these creatures and why we need to study them more. Take, for example, mosquitoes from the genera *Malaya* which live in Africa, the Orient and Australasia, and have a uniquely fluffy and bent proboscis, which is swollen at the end and folded beneath the body when the mosquito is at rest. This unusual feature enables them to get nourishment from a food source no other mosquitoes use: the regurgitated fluid of ants. Yes, they actually choose to eat ant vomit. The adult mosquito flies along until it spies an ant and then it grabs it. It brings the end of its mouthparts into contact with the ant's mouthparts which eventually results in the latter being prised apart and the ant producing a droplet of fluid. This is rapidly processed by the mosquito and the ant is released relatively unharmed! This is the kind of story that reminds me how much I love flies.

Even when we know what flies feed on, crazy things occur. The vinegar fly, *Drosophila flavohirta*, is a pollen feeder but not one we are

The enigmatic-looking male bone skipper fly, *Thyreophora cynophila*. It comes out at night so avoids being caught, which may explain why it was thought to be extinct.

The *Malaya* species of fly with its uniquely fluffy and bent proboscis, which helps when you're feeding off ant vomit.

too familiar with. They were originally from Australia but have been rapidly spreading across the world, presumably within their larval hosts, the eucalypts and *Syzygium* of the myrtle family, where they spend most of their lives. The larvae are considerable pests on other plants and in their introduced range, especially South Africa, have been shown to seriously diminish pollen production and so decrease honey yields. The adults demonstrate a unique feeding method usually only seen in butterflies and moths. They gather up pollen into little balls and vibrate the moistened packages on the upper surface of their proboscis. On analysis of their guts, no pollen is found and so it is assumed they just absorb the broken down nutrients externally. Once more the adaptability of flies, utilising new hosts and novel feeding mechanisms, has resulted in changes to our ecosystems requiring much more research.

They are not the only species to have bizarre tastes. *Megaselia scalaris* – a phorid fly we have discussed before – has also been found eating boot polish and emulsion paint! We are fully aware of how diverse this family is when it comes to utilizing different food sources, but this shows how they are particularly adaptable to living alongside humans. And just when you thought that this incredible family could not get any weirder, a paper published in 2015 by Mexican arachnologist Salima Machkour-M'Rabet and colleagues describes how they found more than 500 larvae living and feeding inside a tarantula. Most dipteran parasitoids reach, at most, double figures of larvae per host – this level of infestation is more common amongst the parasitic wasps! But the larvae are very small and the host is very large – I guess we should be grateful that they have no preference for humans, yet. Though the genetic plasticity of flies to evolve in new environments has been studied, there still remain many unanswered questions.

Melanderia mandibulata is a coastal species of Dolichopodidae found only in California that has a rather unusual morphology. To date they are probably the only species of fly we know that truly is a biting fly – the ends of its labium are strongly sclerotized (hardened), resembling a jaw and functioning in a similar way. However, on consulting the literature and the experts in the doli world, I couldn't find any information about what this species of fly feeds on! For a fly to be so unusual and still not studied is sadly not uncommon.

Not all flies are so shy about coming into contact with humans. Some of our closest companions, the common house flies, are so good at following us that they have now been recorded at the Mount Everest base camp – that is an altitude of 5,364 m (17,598 ft)! The love that this species has for us (and our waste) may turn into a real problem, however. A 2005 publication by a UK researcher Dave Goulson and colleagues determined that, due to climate change, the world population of house flies could increase by 244% by 2080!

That's a lot of flies, even for my tastes. We don't know what is going to happen with climate change and its impact on humans and our environment, but it's clear that flies are integral.

Not all densities of flies are increasing – sadly many are decreasing, and rapidly. Many species found living on isolated islands, sub-Antarctica or at very high altitudes have lost their ability to fly. The lack of predator species in particular enabled more open and accessible habitats to become available and so there was no need for flight. But many of these species are now endangered or have already disappeared. One species of Dolichopididae, *Campsicnemus mirabilis,* was once said to be common on Tantalus Peak, Hawaii. But this once desolate volcanic spot is now crowded both with humans, who have dramatically changed the habitat, and the introduced predacious *Pheidole* ant species, which are predators of these flies. This species is now presumed extinct as it has not been seen since the 1980s. Species loss is a major problem, and species that have yet to be described may, sadly, only receive the honour of being recognized after they are extinct.

Amongst this gloom there is light, with the rediscovery of once-thought extinct species and the discovery of new ones. *Thyreophora cynophila*, commonly called the bone skipper fly, was thought to be the first fly made extinct by humans. It fed on large rotting carcasses which no longer scattered the countryside and it disappeared, we thought, in 1850. But in 2009 it was rediscovered and, maybe now we can determine whether the male's head (it's bright orange) does indeed glow in the dark, as has been claimed, as part of his mating ritual. But there's a hell of a lot of work still to do. A brilliant but scary paper published in 2016 by Paul Hebert and colleagues at the Centre for Biodiversity Genomics, University of Guelph, Canada, details a study they conducted using DNA barcoding techniques – they are arguably the world's leading group when it comes to DNA barcoding – to estimate patterns of species numbers. Both hymenoptera and diptera showed up as being unexpectedly diverse, unexpected given there are only a relatively small number of diptera that have been described in comparison with beetles, for example, which have many more described species. One family they focused on was Cecidomyiidae – the gall midges – which are impossible in most cases to describe from morphological characteristics. Globally around 6,000 species have been recorded but their study estimated that in Canada alone there were 16,000 species. From this figure they extrapolated the global species richness and estimated (based on Canada having 1% of the world fauna) that there were 1.8 million species just in this family. Single-handedly just this one family of flies may eclipse all the species of beetles.

Enormous figures like this give me heart palpitations though, and we won't know how realistic they are for some time to come. The curator in me worries about storage. Will we have enough room for all these new species? Even though we are now just looking at DNA we still need the reference specimens. The ecologist in me also wonders about the impact of describing all these species. We would know about new species names, but what about how they live and breed, what about the impacts both good and bad they might have for the planet? The amount of data we'd be looking at would be vast.

For now, the world of flies is still a relatively unknown one. When asked why I love them so much I reply because no other group is more adaptive, crazy or more ingenious in their morphology and general bad-ass behaviour. Ending with a final silly fact, one species of soldier fly has the longest name on the planet, *Parastratiosphecomyia stratiosphecomyioides,* which translates to 'near soldier wasp-fly wasp-fly like'. If you look it up you can find out how to pronounce it but little else. How can we not want to know more about this species and all the other intriguing species of flies that are out there.

Further Reading

INTRODUCTION

Balashov, Y.S. (1984), Interaction between blood-sucking arthropods and their hosts, and its influence on vector potential. *Ann. Rev. Entom.*, 29: 137–156.

Borkent, A. & Spinelli, G.R. (2007), Neotropical Ceratopogonidae (Diptera, Insecta): Ceratopogonidae. *Ser. Aquat. Biodivers. Latin America*, 4, 198 pp.

Brown, B.V. et al. (2010), *Manual of Central American Diptera, Volume 1*. NRC Research Press, Ottawa, 714 pp.

Brown, B.V. et al. (2010), *Manual of Central American Diptera, Volume 2*. NRC Research Press, Ottawa, 728 pp.

Disney, R.H.L. (1994), *Scuttle Flies: The Phoridae*. Chapman & Hall, London.

Hering, E.M. (1951), *Biology of the Leaf Miners*. W. Junk, The Hague, 420 pp.

Markow, T.A. & O'Grady, P.M. (2006), *Drosophila*. Elsevier, London, 259 pp.

Marshall, S.A. (2012), *Flies: The Natural History and Diversity of Diptera*. Firefly Press Ltd., 616 pp.

Oldroyd, H (1966), *The Natural History of Flies*. W.W. Norton and Co., 372 pp.

Pape, T. et al. (eds.), (2009), *Diptera Diversity: Status, Challenges, and Tools*. Brill Academic, Leiden, 459 pp.

Skidmore, P. (1985), The biology of the Muscidae of the world. *Ser., Entomol.*, 29, xiv, 550 pp.

Yeates, D.K. & Wiegmann, B.M. (2005) *The Evolutionary Biology of Flies*. Columbia University Press, 440 pp.

CHAPTER ONE – THE IMMATURE ONES

Attardo, G.M. et al. (2008), Analysis of milk gland structure and function in *Glossina morsitans*: milk protein production, symbiont populations and fecundity. *J. Insect Physiol.*, 54(8): 1236–1242.

Attardo, G. et al. (2014), Genome sequence of the tsetse fly (*Glossina morsitans*): vector of African trypanosomiasis. *Science*, 344 (6182): 380–386.

Byrne, K. & Nichols, R.A. (1999), *Culex pipiens* in London Underground tunnels: differentiation between surface and subterranean populations. *Heredity*, 82: 7–15.

Li, Y. et al. (2013), A new species of *Ocydromia* Meigen from China, with a key to species from the Palaearctic and Oriental Regions (Diptera, Empidoidea, Ocydromiinae). *ZooKeys*, 349: 1–9.

Sukontason, K. et al. (2004), Ultrastructure of eggshell of *Chrysomya nigripes* Aubertin (Diptera: Calliphoridae). *Parasitol. Res.*, 93(2):151–154.

CHAPTER TWO - THE POLLINATORS

Goldblatt, P. et al. (2004), Pollination by fungus gnats (Diptera: Mycetophilidae) and self-recognition sites in *Tolmiea menziesii* (Saxifragaceae). *Plant Syst. Evol.*, 244: 55–67.

Holloway, B.A. (1976), Pollen-feeding in hover-flies (Diptera: Syrphidae). *New Zeal. J. Zool.*, 3:4, 339–350.

Karolyi, F. et al. (2013), Time management and nectar flow: flower handling and suction feeding in long-proboscid flies (Nemestrinidae: Prosoeca). *Naturwissenschaften*, 100 (11): 1083–1093.

Karolyi, F. et al. (2014), One proboscis, two tasks: adaptations to blood-feeding and nectar-extracting in long-proboscid horse flies (Tabanidae, Philoliche). *Arthropod Struct. Dev.*, 43(5): 403–413.

Orford, K.A. et al. (2015), The forgotten flies: the importance of non-syrphid Diptera as pollinators. *Proc. R. Soc. Lond., B*, 282: 1805.

Potts, S. et al. (2013), *Sustainable pollination services for UK crops*. http://www.reading.ac.uk/caer/Project_IPI_Crops/project_ipi_crops_index.html.

Ssymank, A. et al. (2008), Pollinating flies (Diptera): a major contribution to plant diversity and agricultural production. *Biodivers.*, 9, 86–89.

Tiusanen, M. et al. (2016), One fly to rule them all – muscid flies are the key pollinators in the Arctic. *Proc. R. Soc. Lond., B*, 283: 20161271.

CHAPTER THREE – THE DETRITIVORES

Akers, A.A. (1996), Chapter 19: Adapted to greatest depth. In: *Book of Insects*. University of Florida.

O'Connor, T.K. et al. (2014), Microbial interactions and the ecology and evolution of Hawaiian Drosophilidae. *Front Microbiol.*, 18 (5): 616.

Tamura, K. et al. (1995), Origin of Hawaiian drosophilids inferred from alcohol dehydrogenase gene sequences. In: *Current Topics in Molecular Evolution*, (eds. M. Nei and N Takahata). Pennsylvania State University, pp. 9–18.

Wihlm, M.W. & Courtney, G.W. (2011), The distribution and life history of *Axymyia furcata* McAtee (Diptera: Axymyiidae), a wood inhabiting, semi-aquatic fly. *Proc. Entomol. Soc. Washington* 113(3):385–398.

CHAPTER FOUR – THE COPROPHAGES

Alltech (2015), *2015 Global Feed Survey*. http://www.alltech.com/sites/default/files/global-feed-survey-2015.pdf.

Bernasconi, M.V. et al. (2000), Phylogeny of the Scathophagidae (Diptera, Calyptratae) based on mitochondrial DNA sequences. *Mol. Phylogenet. Evol.*, 16(2): 308–315.

Danovich, T. (2014), *What To Do With All of the Poo*? http://modernfarmer.com/2014/08/manure-usa/.

Emerson, P.M. & Bailey, R.L. (1999), Trachoma and fly control. *Community Eye Health*, 12(32): 57.

Gillieson, D. (2009), *Caves: Processes, Development and Management*. Blackwell Publishing, Malden, 324 pp.

Gleeson, D.M. et al. (2000), The phylogenetic position of the New Zealand batfly, *Mystacinobia zelandica* (Mystacinobiidae; Oestroidea) inferred from mitochondrial 16S ribosomal DNA sequence data. *J. R. Soc. New Zeal.*, 30(2): 155–168.

Holloway, B.A. (1977), A new bat-fly family from New Zealand (Diptera : Mystacinobiidae). *New Zeal. J. Zool.*, 3(4): 313–325.

McAlpine, D.K., (2007), Review of the Borboroidini or Wombat Flies (Diptera: Heteromyzidae) with reconsideration of the status of families Heleomyzidae and Sphaeroceridae, and descriptions of femoral gland-baskets. *Rec. Australian Museum*, 59(3): 143–219.

Petersson, E. & Sivinski , J. (1996), Attraction of a kleptoparasitic sphaerocerid fly (*Norrbomia frigipennis*) to dung beetles (*Phanaeus* spp. and *Canthon* sp.). *J. Insect Behav.*, 9(5): 695–708.

Rozendaal, J.A. (1997), Chapter 6: Houseflies. In:*Vector Control: Methods for Use by Individuals and Communities*. WHO Publications, pp. 302–323.

Sivinski, J., Marshall, S. & Peterson, E. (1999), Kleptoparasitism and phoresy in the Diptera. *The Florida Entomol.*, 82: 179–197.

Unger, K. (2014), *Farm 432: Insect Breeding*. http://www.kunger.at/161540/1591397/overview/farm-432-insect-breeding.

CHAPTER FIVE – THE NECROPHAGES

Batzer, D.P. & Sharitz, R.R. (eds.) (2007), *Ecology of Freshwater and Estuarine Wetlands*. Univ. Calif. Press.

Robinson, W. H. (2005), *Urban Insects and Arachnids: a Handbook of Urban Entomology*. Cam. Univ. Press, 490 pp.

Benecke, M. (2008), Brief survey of the history of forensic entomology. *Acta Biol. Benrodis*,14: 15–38.

Bexfield, A. et al. (2004), Detection and partial characterisation of two antibacterial factors from the excretions/secretions of the medicinal maggot *Lucilia sericata* and their activity against methicillin-resistant *Staphylococcus aureus* (MRSA). *Microbes Infect.*, 6(14):1297–304.

Bhadra, P. et al. (2014), Factors affecting accessibility of bodies disposed in suitcases to blowflies. *Forensic Sci. Int.*, 239: 62–72.

Bonduriansky, R. & Brooks, R.J. (1998), Copulation and oviposition behaviour of *Protopiophila litigata* (Diptera: Piophilidae). *Can. Entomol.*, 130(4): 399–405.

Cannings, R.A. (2012), *Dronefly or rat-tailed maggot (Diptera; Syrphidae)*. http://www.guelphlabservices.com/files/PDC/071DroneFly.pdf.

Carles-Tolra, M. & Prado e Castro, C. (2011), Some dipterans collected on pig carcasses in Portugal Diptera Carnidae, Heleomyzidae, Lauxaniidae and Sphaeroceridae. *Bol. SEA*, 48: 233–236.

Čičková, H. et al. (2012), Biodegradation of pig manure by the housefly, *Musca domestica*: a viable ecological strategy for pig manure management. *PLoS ONE*, 7(3): e32798. doi:10.1371/journal.pone.0032798.

Dowding, V.M. (1967), The function and ecological significance of the pharyngeal ridges occurring in the larvae of some cyclorrhaphous Diptera. *Parasitol.*, 57: 371–388.

Greenberg, B. (1973), *Flies and Disease, Vol. 2: Biology and Disease*. Princeton University Press, NJ, xii + 447 pp.

Marshall, S.A. (1983), *Ceroptera sivinskii*, a new species of Sphaeroceridae (Diptera), In: A genus new to North America, associated with scarab beetles in Southwestern United States. *Proc. Entomol. Soc. Washington*, 85:139–143.

Martin-Vega, D. et al. (2011), The 'coffin fly' *Conicera tibialis* (Diptera: Phoridae) breeding on buried human remains after a postmortem interval of 18 years. *J. Forensic Sci.*, 56: 1654–1656.

McAlpine, D.K. (2011), Review of the Borboroidini or Wombat Flies (Diptera: Heteromyzidae), with reconsideration of the status of families Heleomyzidae and Sphaeroceridae, and descriptions of femoral gland-baskets. *Rec. Australian Museum*, 59 (3): 143–219.

Miller, P.L. (1984), Alternative reproductive routines in a small fly, *Puliciphora borinquenensis* (Diptera: Phoridae). *Ecol. Entomol.*, 9(3): 293–302.

Özsisli, T. & Disney R.H.L. (2011), First records for Turkish fauna: *Megaselia brevissima* (Schmitz, 1924) and *Megaselia scalaris* (Loew, 1866) (Diptera: Phoridae). *Türk Entomol. Bült*, 1: 31–33.

Thomas, S. (2010), *Surgical Dressings and Wound Management*. Medetec, Cardiff, 778 pp. http://www.sea-entomologia.org/PDF/001007BSEA46Thyreophorabr.pdf.

CHAPTER SIX – THE VEGETARIANS

Badii, K.B. et al. (2015), Review of the pest status, economic impact and management of fruit-infesting flies (Diptera: Tephritidae) in Africa. *Afr., J. Agric. Res.*, 10(12): 1488–1498.

Camazine, S. (1985), Leaping locomotion in *Mycetophila cingulum* (Diptera: Mycetophilidae): prepupation dispersal mechanism. *Ann. Entomol. Soc. Am.*, 79(1):140–145.

de Bruijn, F.J. (2015), *Biological Nitrogen Fixation*. Wiley-Blackwell Publishers, pp. 1–1196.

Felt, E.P. (1918), Gall insects and their relations to plants. *Sci. Monthly*, 6(6): 509–525.

Heads, P.A. & Lawton, J.H. (1983), Studies on the natural enemy complex of the holly leaf-miner: the effects of scale on the detection of aggregative responses and the implications for biological control. *Oikos*, 40(2): 267–276.

Hutson, A.M. et al. (1980), Mycetophilidae (Bolitophilinae, Ditomyiinae, Diadocidiinae, Keroplatinae, Sciophilinae and Manotinae) Diptera, Nematocera. *Handbooks for the Identification of British Insects, Vol. IX, Part 3*.

Katayama, N. et al. (2014), Sexual selection on wing interference patterns in *Drosophila melanogaster*. *PNAS*, 111(42): 15144–15148.

Shevtsova, E. et al. (2011), Stable structural color patterns displayed on transparent insect wings. *PNAS*, 108(2): 668–673.

CHAPTER SEVEN – THE FUNGIVORES

Broadhead, E.C. (1984), Adaptations for fungal grazing in Lauxaniid flies. *J. Nat. Hist.*, 8:639–649.

Chandler, P.J. (2001), The flat-footed flies: (Diptera: Opetiidae and Platypezidae) of Europe. *Fauna Entomol. Scand.*, 36, 276 pp.

Colless, D.H. (1962), A new Australian genus and family of Diptera (Nematocera: Perissommatidae). *Australian J. Zool.*, 10(3): 519–536.

Colless, D.H. (1969), The genus *Perissomma* (Diptera: Perissommatidae) with new species from Australia and Chile. *Australian J. Zool.*, 17(4): 719–728.

Hackman, W. & Meinander, M. (1979), Diptera feeding as larvae on macrofungi in Finland. *Ann. Zool. Fennici*, 16(1): 50–83.

Hippa, H. et al. (2005), New taxa of the Lygistorrhinidae (Diptera: Sciaroidea) and their implications for a phylogenetic analysis of the family. *Zootaxa*, 960: 1–34.

Hippa, H. et al. (2009), Review of the genus *Nepaletricha* Chandler (Diptera, Rangomaramidae), with description of new species from Thailand and Vietnam. *Zootaxa*, 2174: 18–26.

Lewandowski, M. et al. (2012), Biology and morphometry of *Megaselia halterata*, an important insect pest of mushrooms. *Bull. Insect.*, 65:1–8.

McAlpine, D.K. (1973), Observations on sexual behaviour in some Australian Platystomatidae (Diptera, Schizophora). *Rec. Australian Museum*, 29(1): 1–10.

Rindal, E. & Gammelmo, Ø. (2007), On the family Diadocidiidae (Diptera, Sciaroidea) in Norway. *Norw. J. Entomol.*, 54: 69–74.

Rohacek, J. (1999), A revision and re-classification of the genus *Paranthomyza* Czerny, with description of a new genus of Anthomyzidae (Diptera). *Studia Dipterologica*, 6(2): 239–270.

CHAPTER EIGHT – THE PREDATORS

Berg, C.O. & Knutson, L. (1978), Biology and systematics of the Sciomyzidae. *Ann. Rev. Entomol.*, 23: 239–258.

Burger, J.F. et al. (1980), The habits and life history of *Oedoparena glauca* (Diptera: Dryomyzidae), a predator of barnacles. *Proc. Entomol. Soc. Wash.*, 82: 360–377.

Cregan, M. B. (1941), *Generic relationships of the Dolichopodidae (Diptera) Based on a Study of the Mouthparts*. Urbana, University of Illinois Press.

Cumming, J.M. (1994), Sexual selection and the evolution of dance fly mating systems (Diptera: Empididae; Empidinae). *Can. Entomol.*, 126: 907–920.

Davis, C.J. et al. (1961), Introduction of the liver fluke snail predator, *Sciomyza dorsata* (Sciomyzidae, Diptera), in Hawaii. *Proc. Hawaiian Entomol. Soc.*, 17:395–397.

Davis, C.J. & Krauss, N.H.L. (1962), Recent Introductions for Biological Control in Hawaii. *Proc. Hawaiian Entomol. Soc.*, 18: 125–127.

Downes, J.A. (1978), Feeding and mating in the insectivorous Ceratopogoninae (Diptera). *Mem. Entomol. Soc. Can.*, 104: 1–62.

Fry, B.G. at al. (2009), The toxicogenomic multiverse: convergent recruitment of proteins into animal venoms. *Ann. Rev. Genomics Hum. Genet.*, 10:483–511.

Germann C. et al. (2010), Legs of deception: disagreement between molecular markers and morphology of long-legged flies (Diptera, Dolichopodidae). *J. Zool. System. Evol. Res.*, 48: 238–247.

Hurley, R.L. & Runyon, J.B. (2009), A review of *Erebomyia* (Diptera: Dolichopodidae), with descriptions of three new species. *Zootaxa*, 2054, 38–48.

Land, M.F. (1993), Chasing and pursuit in the dolichopodid fly *Poecilobothrus nobilitatus*. *J. Comp. Physiol. A*, 173 (5): 605–613.

Menin, M. & Giaretta, A.A. (2003), Predation on foam nests of leptodactyline frogs (Anura: Leptodactylidae) by larvae of *Beckeriella niger* (Diptera: Ephydridae). *J. Zool.*, 261: 239–243.

Oba, Y. et al. (2011), The terrestrial bioluminescent animals of Japan. *Zool. Sci.*, 28:771–789.

Piper, R. (2007), *Extraordinary Animals: An Encyclopaedia of Curious and Unusual Animals*. Greenwood Publishing, 321 pp.

Sadowski, J.A. et al. (1999), The evolution of empty nuptial gifts in a dance fly, (Diptera: Empididae): bigger isn't always better. *Behav. Ecol. Sociobiol.*, 1999:161–166.

Satô, M. (1991), Comparative morphology of the mouthparts of the family Dolichopodidae (Diptera).*Insecta Matsumurana*, 45: 49–75.

von Reumont, B.M. et al. (2014), Quo vadis venomics? A roadmap to neglected venomous invertebrates. *Toxins*, 6: 3488–3551.

Zimmer, M. et al. (2003), Courtship in long-legged flies (Diptera: Dolichopodidae): function and evolution of signals. *Behav. Ecol.*, 14: 526–530.

CHAPTER TEN – THE PARASITES

Brauer, F (1885), Systematisch-zoologische Studien. *Sber. Akad. Wiss. Wien*, 1(91): 237–413, 1 pl.

Calhau, J. et al. (2014), Taxonomic revision of *Pseudorhopalia* Wilcox & Papavero, 1971 (Insecta, Diptera, Mydidae, Rhopaliinae), with description of a new species from the Brazilian. *Zootaxa*, 3884 (4): 333–346.

Coupland, J. & Barker, G.M. (2004), Diptera as predators and parasitoids of terrestrial gastropods, with emphasis on Phoridae, Calliphoridae, Sarcophagidae, Muscidae and Fanniidae, In: *Natural Enemies of Terrestrial Molluscs*, Barker, G.M (ed). CABI, Cambridge, pp. 85–154.

Feener, D.H. & Brown, B.V. (1997), Diptera as parasitoids. *Ann. Rev. Entomol.*, 42: 73–97.

Fessl, B. et al. (2006), The life-cycle of *Philornis downsi* (Diptera: Muscidae) parasitizing Darwin's finches and its impacts on nestling survival. *Parasitol.*, 133(6):739–47.

Halbert, S.E. (2008), *Tri-Ology Report: Entomology section: Arthropod Detection*. FDACS-Div. Plant Industry.

Koenig, D.P. & Young, C.W. (2007), First observation of parasitic relations between big-headed flies, *Nephrocerus* Zetterstedt (Diptera: Pipunculidae) and crane flies, *Tipula* Linnaeus (Diptera: Tipulidae: Tipulinae), with larval and puparial descriptions for the genus *Nephrocerus*. *Proc. Entomol. Soc. Wash.*, 109: 52–65.

Land, M.F. (1993), The visual control of courtship behaviour in the fly *Poecilobothrus nobilitatus*. *J. Comp. Physiol. A*, 173: 595–503.

Schmidt, J.O. (1982), Biochemistry of insect venoms. *Ann. Rev. Entomol.*, 27:339–68.

Toft, S. et al. (2012), Parasitoid suppression and life-history modifications in a wolf spider following infection by larvae of an acrocerid fly. *J. Arachnol.*, 40(1):13–17.

Wardlaw, J.C. et al. (2000), Observations on the life cycle of Medina collaris (Fallén) (Dipt., Tachinidae). *Entomol. Monthly Mag.*, 136: 21–29.

Zayed, A.A. (1998), Localization and migration route of *Cephalopina titillator* (Diptera: Oestridae) larvae in the head of infested camels (*Camelus dromedarius*). *Vet. Parasitol.*, 80(1): 65–70.

CHAPTER ELEVEN – THE SANGUIVORES

Abbitt, B. & Abbitt, L.G. (1982), Fatal exsanguination of cattle attributed to an attack of salt marsh mosquitoes (*Aedes sollicitans*). *J. Am. Vet. Med. Assoc.*, 179(12):1397–400.

Benoit, J.B. et al. (2011), Drinking a hot meal elicits a protective heat shock response in mosquitoes. *Proc. Nat. Acad. Sci., USA*, 108 (19): 8026–8029.

Bowman, D.D. (1985), *Georgis' Parasitology for Veterinarians*. Elsevier, 496 pp.

Braack, L. & Pont, A.C. (2012), Rediscovery of *Haematobosca zuluensis* (Zumpt), (Diptera, Stomoxyinae): re-description and amended keys for the genus. *Parasites and Vectors*, 5(267): 1–7.

Burger, J.F. & Anderson, J.R. (1974), Taxonomy and life history of the moose fly, *Haematobosca alcis*, and its association with the moose, *Alces alces shirasi* in Yellowstone National Park. *Ann. Entomol. Soc. Am.*, 67: 204–214.

Camp, J.V. (2006), *Host Attraction and Host Selection in the Family Corethrellidae (Wood And Borkent) (Diptera)*. Electronic Theses & Dissertations. Paper 728.

Downes, J.A. (1966), Observations on the mating behaviour of the crab hole mosquito *Deinocerites cancer* (Diptera: Culicidae). *Can. Entomol.*, 98(11):1169–1177.

FAO, United Nations (2013), *Edible Insects: future prospects for food and feed security*. FAO Forestry Paper, 208 pp.

Hancock, R.G. et al. (1990), Tests of *Sabethes cyaneus* leg paddle function in mating and flight. *J. Am. Mosq. Control Assoc.*, 6(4):733–5.

Lahondare, C. & Lazzari, C.R. (2012), Mosquitoes cool down during blood feeding to avoid overheating. *Current Biol.*, 22 (1): 40–45.

Mullen, G.R. & Durden, L.A. (2002), *Medical and Veterinary Entomology*. Academic Press, New York. 597 pp.

Noirti, N.C. & Boiteu, X.P. (1979), Death of 25 farm animals (including 24 cattle) attributed to the bites of Simuliidae (black flies) in the Vosges. *Bull. mens. Soc. vét. prat. Fr.*, 63: 41–54.

Noirtin, C. et al. (1981), Les simulies, nuisance pour le bétail dans les Vosges : les origines de leur pullulation et les méthodes de lutte. Cahiers ORSTOM. *Sér. Entomol. Médicale et Parasitol.*, 19(2): 101–112.

Pollock, J.N. (1982), *Training Manual for Tsetse Control Personnel, Vol. 1: Tsetse biology, systematics and distribution, techniques*. FAO, Rome.

Provost, M.W. & Haeger, J.S. (1967), Mating and pupal attendance in *Deinocerites cancer* and comparisons with *Opifex fuscus* (Diptera: Culicidae). *Ann. Entomol. Soc. Am.*, 60: 565–574.

Rozendaal, J.A. (1997), Chapter 2, Tsetse flies. In: *Vector Control: Methods for Use by Individuals and Communities*. WHO, pp 178–192.

Sallum, M.A.M. & Flores, D.C. (2004), Ultrastructure of the eggs of two species of *Anopheles* (Anopheles) Meigen (Diptera, Culicidae). *Rev. Bras. Entomol.*, 48(2):185–192.

Thorpe, W. H. (1942), Observations on *Stomoxys ochrosoma* Speiser (Diptera Muscidae) as an associate of army ants (Dorylinae) in East Africa. *Proc. R. Entomol. Soc. Lond. A*, 17: 38–41.

Walker, A.R. (1994), *Arthropods of Humans and Domestic Animals: A Guide to Preliminary Identification*. Springer, The Netherlands, 214 pp.

Wirth, W.W. (1972), Midges sucking blood of caterpillars (Diptera: Ceratopogonidae). *J. Lepid. Soc.*, 26: 65.

Wirth, W.W. (1975), Biological notes and new synonymy in *Forcipomyia* (Diptera: Ceratopogonidae). *Florida Entomol.*, 58(4): 243–245.

Wirth, W.W. (1994), The subgenus *Atrichopogon* (*Lophomyidium*) with a revision of the Nearctic species (Diptera: Ceratopogonidae). *Insecta Mundi*, 8:17–36.

THE END

Hebert, P.D.N. et al. (2016), Counting animal species with DNA barcodes. *Can. Insects. Phil. Trans. R. Soc. B*, 371.

Goulson, D. et al. (2005), Predicting calyptrate fly populations from the weather, and probable consequences of climate change. *J. App. Ecol.*, 42, 795–804.

Machjour-M'Rabet, S. et al. (2015), *Megaselia scalaris* (Diptera: Phoridae): an opportunistic endoparasitoid of the endangered Mexican redrump tarantula, *Brachypelma vagans* (Araneae: Theraphosidae). *J. Arachnol.*, 43(1): 115–119.7.

Index

Picture credits

p.9 © Laszlo Ilyes/Wikimedia Commons; p.11, 171 ©Solvin Zankl/naturepl.com; p.26 © Kim Taylor/naturepl.com; p.28 © Maria Anice Mureb Sallum; © Stephen L. Doggett; p.33 © Dr. Chen W. Young (Carnegie Museum); p.37 Arlo Pelegrin; p. 47, 72 ©Steven Falk; p.50 ©Australian Museum, Sydney; p.59 © C T Johansson/Wikimedia Commons; p.60 © Neil Lucas/naturepl.com; p. 62, 76 ©Piotr Naskrecki; p.66 © Xavier Vázquez/Wikimedia Commons; p. 69 © Kevin Mackenzie, University of Aberdeen, Wellcome Images; p.77 © Mark W. Moffett, Minden Pictures; p.88 ©Robin Bailey; p.90 © Seth Ausubel; p.101 © Amir Weinstein; p. 104 ©Eye of Science/Science Photo Library; p. 114 ©Rod Williams/naturepl.com; p. 116 © Stephen Luk; p.117 © R. Bonduriansky; p. 119 © Edward L. Ruden; p.122 © Susan Wineriter, USDA Agricultural Research Service, Bugwood.org; p.128 ©Smithsonian Libraries; p.133 © Alvesgaspar/Wikimedia Commons; p.139 ©Scott Bauer/US Department of Agriculture/Science Photo Library; p.142 © Karsten Sund, NHM, Oslo, Norway; p. 148, 200 © University of Nebraska Department of Entomology; p. 149 ©Peter Kerr; p. 154 ©Graham Wise; p.155 © Gaimari, S.D., & V.C. Silva. 2010. Revision of the Neotropical subfamily Eurychoromyiinae (Diptera: Lauxaniidae). Zootaxa 2342: 1-64.), and specifically that it is Figure 5E, reproduced with permission of the first author; p.157 ©Tony Daley; p.173 © Greg W. Lasley ALL RIGHTS RESERVED; p.176 bottom right © Astrobradley/Wikimedia Commons; p.184, 230 ©Tom Murray; p.187 ©Michael Bell; p.189 © Márcia Couri; p.191 © Pavel Kirillov/Wikimedia Commons; p.202 ©Wendy Porras; p.203 © United States Department of Agriculture, Agricultural Research Service; p.205 ©Joyce Gross; p.210 ©Miles Zhang; p.217 top ©Steven Falk; p217 bottom © Magne Flåten; p.219 © Gayle and Jeanell Strickland; p.220, 221 © A. W. Thomas, Ph.D.; p. 222 © Scott Bauer; p.225 © G. Kunz; p.228 ©Paul Bertner; p.233 © Suzanne Koptur; p.234 © Ma et al.; licensee BioMed Central Ltd. 2013; p.238 © J. Stoffer, Walter Reed Biosystematics Unit.

Unless otherwise stated images copyright of Natural History Museum, London.

Every effort has been made to contact and accurately credit all copyright holders. If we have been unsuccessful, we apologise and welcome correction for future editions and reprints.

Acknowledgements

Firstly, I would like to thank my editors, reviewers, imagers and everyone who was involved in donating material, stories and research to create this book – you took a jumble of excitable anecdotes and crafted them into something readable. Apologies to the picture editor for having to source so many "disturbing" pictures (sadly most couldn't go in).

I would like to thank the Natural History Museum, London and all its staff (especially 'team Diptera') – the enthusiasm from both is one of the reasons I put pen to paper. A massive thank you to the Dipterists Forum, whose members have taught me so much and who provide so much more in terms of time and expertise.

I would like to thank all my friends (but especially Ruth, Rusc, Nicky, Polly and Sarah) and my family, who for years have put up with disgusting conversations about flies, at generally the most inappropriate times, and so really are to blame for this book happening.

And most of all I would like to thank my Mum and Dave – yes; you are both rude about flies most of the time, but luckily for me, critical editors of the written word.